5G Non-Terrestrial Networks

IEEE Press
445 Hoes Lane
Piscataway, NJ 08854

5G Non-Terrestrial Networks

Technologies, Standards, and System Design

Alessandro Vanelli-Coralli
University of Bologna
Italy

Nicolas Chuberre
Thales Alenia Space – France
France

Gino Masini
Ericsson AB
Sweden

Alessandro Guidotti
CNIT, Research Unit at the University of Bologna
Italy

Mohamed El Jaafari
Thales Alenia Space – France
France

IEEE PRESS

WILEY

Published by John Wiley & Sons, Inc., Hoboken, New Jersey.
Published simultaneously in Canada.

For general information on our other products and services or for technical support, please contact our Customer Care Department within the United States at (800) 762-2974, outside the United States at (317) 572-3993 or fax (317) 572-4002.

Wiley also publishes its books in a variety of electronic formats. Some content that appears in print may not be available in electronic formats. For more information about Wiley products, visit our web site at www.wiley.com.

Library of Congress Cataloging-in-Publication Data:

Names: Vanelli-Coralli, Alessandro, author.
Title: 5G non-terrestrial networks / Alessandro Vanelli-Coralli [and four others].
Description: Hoboken, New Jersey : Wiley, [2024] | Includes index.
Identifiers: LCCN 2023048993 (print) | LCCN 2023048994 (ebook) | ISBN 9781119891154 (hardback) | ISBN 9781119891161 (adobe pdf) | ISBN 9781119891178 (epub)
Subjects: LCSH: 5G mobile communication systems. | Artificial satellites.
Classification: LCC TK5103.25 .V36 2024 (print) | LCC TK5103.25 (ebook) | DDC 621.3845/6–dc23/eng/20231122
LC record available at https://lccn.loc.gov/2023048993
LC ebook record available at https://lccn.loc.gov/2023048994

Cover Design: Wiley
Cover Image: © metamorworks/Shutterstock

Set in 9.5/12.5pt STIXTwoText by Straive, Chennai, India
SKY10065688_012324

To our families, Elena, Leonardo, Caterina, Daniela, Luigi, Lucia, Giulia, Nabiha, Rassane, Maryam, Anne-Claire, Marine, Alix, Cyprien, Xavier, and Hélène, for their continuous support and encouragement.

Contents

Preface

For many years, the cellular industry has had the goal to provide its services anytime, anywhere, and on any device (ATAWAD). Such a goal was always postponed to the next generation of mobile systems. What became a dream will now be reality with 3rd Generation Partnership Project (3GPP) Release-17 technical specifications allowing the addition of non-terrestrial network (NTN) components in the 5G system. Leveraging this global standard, industrial initiatives aim at developing and deploying new satellite networks that are able to directly serve smartphones and IoT devices. Moreover, this standard will contribute to create a global market for broadband satellite networks operating in above 10 GHz frequency bands facilitating their interworking with terrestrial communication infrastructure for the support of backhaul services, as well as connectivity to any moving platforms (aircraft, vessels, trains, and land vehicles).

This book aims at providing an overview of this 3GPP-defined "NTN" standard. Aimed at easing the inclusion of NTN elements in the 5G ecosystem, the work was conducted within the 3GPP starting with the identification of the technical challenges and potential solutions to support New Radio (NR) protocols and features over satellite links. This marked a radical paradigm shift in how mobile communication standards are defined, as now we are moving to satellite-terrestrial integration. The first global standard for an NTN component in the 5G system was published mid-2022 in the context of 3GPP Release-17. This standard provides the specifications for 5G systems to support a satellite component. However, it should be mentioned that it does not just represent a set of technical specifications, but it rather opens the door for the satellite industry to enter the 3GPP ecosystem, which involved more than 800 organizations at global level working to ensure a global market for telecommunications.

The 3GPP-defined NTN standard is the result of a massive joint effort between stakeholders from both the terrestrial and satellite industries, which leads to a two-fold benefit. On the one hand, 3GPP can now truly achieve global service continuity and network resiliency; on the other hand, satellite stakeholders can now access the unified and global 3GPP ecosystem and, as such, the possibility to reduce the costs through economy of scale. Moreover, the inclusion of a non-terrestrial component in 3GPP can also yield huge benefits for the satellite industry as ground systems exploiting equipment coming from different providers are now available. Notably, this standard is also supported by vertical stakeholders (including public safety, transportation, and automotive) calling for the seamless integration of satellite and terrestrial components and the support of 5G features across these different radio access technologies.

The addition of the NTN component in 5G systems is already creating a global market for satellite communication industry stakeholders and a unique technology framework for satellite networks based on whatever orbit, frequency band, and service provision considering a variety of terminal types (including smartphones and very small aperture terminals), as well as an open architecture approach.

While the current Release-17 NTN standard provides a solid ground for future satellite networks integrated into the 5G system, a significant innovation breakthrough in technologies, techniques, and architectures is needed to prepare for next-generation satellite networks based on Release-19 and beyond, which will pave the way for 6G communications. At the time of writing this book, 3GPP Release-18 is being defined and it includes enhancing features to further improve the performance or to introduce new capabilities for the support of NTN. Moreover, the discussions on the potential topics to be addressed in the 3GPP Release-19 package are already ongoing.

In this context, this book aims to provide a complete and comprehensive overview of the study and normative activities that led to the definition of the first NTN component in the 3GPP 5G ecosystem within Release-17 and to provide a guideline for the ongoing normative work in 3GPP Release-18. More specifically, the following topics are addressed:

- Chapter 1 provides an introduction to satellite communications and NTNs and what their role is in 5G systems. The evolution of 3GPP specifications throughout the study phase (Release-15 and -16) and the normative phase (Release-17 and beyond) is discussed.
- Chapter 2 introduces the 3GPP ecosystem and the global 5G standard, in terms of architectures (core and radio access), enabling features, and interfaces and protocols. This chapter introduces the reader to 5G systems and 3GPP procedures.
- Chapter 3 provides an overview of NTNs, addressing the network elements (ground, space, and user segment), orbits and orbital propagation aspects, Earth-satellite geometry, and link budget computation. The scope of this chapter allows readers who are not experts in satellite communications to enter the 3GPP NTN world and understand its underlying principles.

- Chapter 4 is dedicated to NR NTN architectures and network protocols, interfaces, and functionalities for both the *user* and *control* planes; these discussions are not limited to Release-17, which is built assuming transparent payloads, but also covers the more advanced options with regenerative payloads that are expected for 5G-Advanced. Aspects related to the interworking with terrestrial network components are also covered.

- Chapter 5 reports an extensive description of the NTN radio interface at physical layer, covering the basic transmission principles (waveform, modulation and coding, multiple access scheme, framing, operating frequency, and radio channels), more advanced topics, and the necessary physical layer mechanisms and procedure modifications for the 5G NR to support a satellite-based access.

- Chapter 6 provides an extensive review of the impacts on the system architectures and network protocols that the introduction of the NTN characteristics has. In particular, the discussion addresses the handling of quality of service (QoS), attachment procedures for the user equipment (UE), mobility, feeder link switch, and network management. As for the previous chapter, the focus is on Release-17, but also the ongoing enhancements for Release-18 are presented.

- Chapter 7 addresses radio frequency (RF) and radio resource management (RRM) aspects, in terms of requirements and target performance, also taking into account the recommendations set forth by the International Telecommunication Union Radiocommunication sector (ITU-R).

- Chapter 8 provides an overview of narrowband Internet-of-Things (NB-IoT) and enhanced Machine Type Communications (eMTC) via NTN, covering 3GPP normative activities and the impact of the NTN channel on radio protocols and the related enhancements.

- Chapter 9 covers the further enhancements and capabilities of NTNs in the evolving context of 5G-Advanced as well as 6G, including both standardization aspects and industrial perspectives. An overview of the preliminary expectations related to NTN in 6G systems is detailed.

December 22, 2023

Alessandro Vanelli-Coralli
University of Bologna
Nicolas Chuberre
Thales Alenia Space
Gino Masini
Ercisson AB
Alessandro Guidotti
CNIT Research Unit at the University of Bologna
Mohamed El Jaafari
Thales Alenia Space

We have

- LinkedIn page: https://www.linkedin.com/company/5g-ntn-book/
- Email: 5gntn.book@gmail.com

Follow us on LinkedIn:

Email
5gntn.book@gmail.com

About the Authors

Alessandro Vanelli-Coralli, PhD, is a full professor of telecommunications at the Department of Electrical, Electronic, and Information Engineering "Guglielmo Marconi," University of Bologna, Italy. Since March 2022, he has been a research fellow in the Digital Integrated Circuits and Systems group at ETH Zurich (CH) working on RISC-V-based multicore platforms for 5G and 6G signal processing. From January to June 2021, he was a visiting professor at ETH Zurich working on Internet of Things (IoT) software-defined radio. In 2003 and 2005, he was a visiting scientist at Qualcomm Inc. (San Diego, CA). His research activity focuses on wireless communication with specific emphasis on 5G and 6G systems and non-terrestrial networks (NTNs). His pioneering work on heterogeneous satellite networks in 2009 set the basis for the concept of 3D multidimensional NTNs (A. Vanelli-Coralli et al. "The ISICOM Architecture," 2009 International Workshop on Satellite and Space Communications). He participates in national and international research projects on satellite mobile communication systems serving as scientific responsible and prime contractor for several European Space Agency (ESA) and European Commission funded projects. He participates in industrial and scientific fora and bodies, he is responsible for the vision and research strategy task force of the NetworldEurope SatCom Working Group, and he is the delegate for the University of Bologna in the 6G-IA, ETSI, and 3GPP. He served as a member of the editorial board of the Wiley *InterScience Journal on Satellite Communications and Networks*, as guest co-editor for several special issues in international scientific journals, and since 2010 he has been the general co-chairman of the IEEE ASMS Conference. He co-authored more than 290 peer-reviewed papers and he is co-recipient of several best paper awards.

He is an IEEE senior member and the 2019 recipient of the IEEE Satellite Communications Technical Recognition Award.

Nicolas Chuberre graduated in 1988 from Ecole Supérieure d'Ingénieur en Electronique et Electrotechnique in Paris. Previously with Nokia and Alcatel Mobile to design signal processing algorithms, medium access control protocols, and test tools for 2G cellular handsets and systems assembly, he then joined Thales Alenia Space to manage the development of satellite payload equipment and the design of advanced satellite communication systems (GEO and non-GEO). He has successfully initiated and led several European collaborative research projects in FP6, FP7, and H2020, as well as the ESA ARTES context. He has been chairing the SatCom Working Group of Networld2020 technology platforms (https://www.networld2020.eu/) for nine years and was a member of the partnership board of the 5G Infrastructure Association (http://5g-ppp.eu/). He has published several papers on innovative satellite system concepts. Currently, he is defining and developing satellite solutions for 5G and 6G systems. In addition, he is the lead representative of Thales in 3GPP TSG RAN where he has been the rapporteur of the standardization on the integration of satellite in 5G since 2017 (https://www.3gpp.org/news-events/partners-news/2254-ntn_rel17). He also chairs since 2006 the satellite communication and navigation working group at ETSI (www.etsi.org). Last, he is the technical manager of the Horizon Europe research project "6G-NTN" (https://www.6g-ntn.eu/).

Gino Masini is principal researcher with Ericsson in Sweden. In his many years in telecommunications in both industry and academia, he has worked with microwave antennas and propagation, satellite telecommunications, microwave circuit research and development, backhaul network design, and radio access network architecture. He received his electronics engineering degree from Politecnico di Milano in 1996 and an MBA from SDA Bocconi School of Management in Milano in 2008. He started as a researcher at Politecnico di Milano, working on projects for the ESA and the Italian Space Agency (ASI); through that activity, he contributed to the background for millimeter wave propagation experiments such as the *Alphasat* satellite mission. He later joined Ericsson, working with antennas and planning for microwave radio links and with MMIC development. Since 2009 he has been working with 4G and 5G radio access network architecture. He has been working in standards since 2001, having attended 3GPP, ETSI, ITU, and CEPT, among others. He has been active in 3GPP since 2009, he was RAN WG3 chairman from 2017 to 2021, overseeing the standardization of 5G radio network architecture, interfaces, and protocols. He is the author of more than 70 patents and several scientific publications, and he is co-author of books on 5G radio access networks; he also holds a "Six Sigma" certification.

Alessandro Guidotti received the Dr. Ing. degree (*magna cum laude*) in telecommunications engineering and the PhD in electronics, computer science, and telecommunications from the University of Bologna (Italy) in 2008 and 2012, respectively. In 2012 he joined the Department of Electrical, Electronic, and Information Engineering (DEI) at the University of Bologna. During his PhD from 2008 to 2012, he was the representative for the Italian Administration within the CEPT SE-43 working group on cognitive radios. In 2011–2012, he spent some months as a visiting researcher at SUPELEC (Paris, France) working on the application of stochastic geometry to interference characterization in wireless networks. From 2012 to 2014 and from 2014 to 2021, he was a post-doctoral researcher and research associate at the University of Bologna, respectively. Since 2021, he has been a senior researcher within the National Inter-University Consortium for Telecommunications (CNIT), working at the research unit of the University of Bologna. He participates in national and international research projects on satellite and terrestrial mobile communication systems. His research interests are in the area of wireless communication systems, NTNs, satellite-terrestrial networks, and digital transmission techniques. He is a member of the IEEE AESS Technical Panel on Glue Technologies for Space Systems, and since 2016 he is TPC and Publication Co-Chair of the IEEE ASMS/SPSC Conference.

Mohamed El Jaafari is a radio access network specialist engineer with more than 23 years of experience in cellular communications, including 5G NR, eUTRAN, GERAN, and cellular IoT. He received an engineering degree in telecommunications from EMI in 1999. He is an expert in radio access network design, radio frequency planning, radio network optimization, and radio access network system dimensioning with extensive multi-vendor experience. He currently conducts extensive research work on 5G NR NTN and IoT NTN. He joined the R&D department of the Telecommunication Business Line of Thales Alenia Space in 2020. He is the lead representative of Thales in 3GPP RAN1 working group where he is a feature lead for the 3GPP work item on satellite integration in 5G. Currently, he is defining and developing solutions for 5G NR to support NTNs. He is active in national and international research projects on wireless and satellite communication systems in several ESA and European Commission funded projects. His research interests include 3GPP wireless communication systems, satellite communications, and their integration in 5G and future 6G networks. He has recently led the drafting of a special edition on NTN standards, https://onlinelibrary.wiley.com/toc/15420981/2023/41/3.

Acknowledgments

The authors would like to express their special gratitude to Mr. Dorin Panaitopol from Thales, for the valuable inputs related to radio frequency (RF) and radio resource management (RRM) aspects and his careful review of the corresponding chapter, and to Dr. Carla Amatetti from the University of Bologna, Italy, for the discussions and contributions on narrowband Internet of Things (NB-IoT) via non-terrestrial network (NTN).

This NTN standardization adventure has involved numerous 3rd Generation Partnership Project (3GPP) experts from the global telecom industry. It is fair to praise the work of 3GPP leadership that had the difficult task of moderating the discussions between two different ecosystems (satellite and mobile industry) in the different groups of 3GPP with an attempt to build consensus on the various aspects of the NTN standard.

- TSG-SA chair: Erik Guttman (Samsung) then Georg Mayer (Huawei) then Puneet Jain (Intel)
- SA1 WG chair: Toon Norp (TNO) then Jose Almodovar (TNO)
- SA2 WG chair: Frank Mademann (Huawei) then Puneet Jain (Intel)
- SA3 WG chair: Suresh Nair (Nokia)
- SA3-LI WG chair: Alex Leadbeater (BT group)
- SA5 WG chair: Thomas Tovinger (Ericsson)
- TSG-RAN chair: Balazs Bertenyi (Nokia) then Wanshi Chen (Qualcomm)
- RAN1 WG chair: Wanshi Chen (Qualcomm) then Younsun Kim (Samsung)
- RAN1 WG vice chair: Havish Koorapaty (Ericsson) then David Mazzarese (Huawei)
- RAN2 WG chair: Richard Burbridge (Intel) then Johan Johansson (Mediatek)
- RAN2 WG vice chairs: Diana Pani (Inter Digital) and Sergio Parolari (ZTE)

- RAN3 WG chair: Gao Yin (ZTE) who took over from Gino Masini (Ericsson)
- RAN4 WG Chair: Steven Chen (Futurewei) then Xizeng Dai (Huawei)
- RAN4 WG vice chairs: Haijie Qiu (Samsung) on RF aspects and Andrey Chervyakov (Intel) on RRM aspects

Special thanks to the many experts involved very early in the NTN standardization work among which are: Tommi Jamsa (Huawei), Frank Hsieh (Nokia), Gilles Charbit (Mediatek), Nan Zhang (ZTE), Xiaofeng Wang (Qualcomm), Philippe Reininger (Huawei), Martin Israelsson (Ericsson), Stefano Cioni (ESA), Munira Jaffar (Hughes), Olof Liberg (Ericsson), Thibaud Deleu (Thales), Baptiste Chamaillard (Thales), Thomas Heyn (Fraunhofer IIS), Thomas Haustein (Fraunhofer HHI), Mohamed El Jaafari (Thales), Cyril Michel (Thales), Relja Djapic (TNO), Saso Stojanovski (Intel), Hannu Hietalahti (Nokia), Matthew Baker (Nokia), Thomas Chapman (Ericsson), Luca Lodigiani (Inmarsat), and Jean-Yves Fine (Thales).

The support of ETSI MCC secretaries such as Joern Krause, Patrick Merias, and Maurice Pope in the various 3GPP groups shall also be underlined.

Current industry activities on 5G via satellite would not have been possible without

- the support of:
  o the European Space Agency (especially Antonio Franchi, Xavier Lobao, Maria Guta, and Dr. Riccardo di Gaudenzi)
  o the European Commission (especially Mr. Bernard Barani)
  o the European Telecommunication Standard Institute (especially Mr. Adrian Scrase)
  o space industry's leaders (Bertrand Maureau/Thales, Didier Le Boulc'h/Thales, Stéphane Anjuere/Thales, Lin-Nan Lee/Hughes Network System)
- And the groundbreaking work by a number of pioneers in our field. Among them:
  o Prof. Francesco Carassa (Politecnico di Milano): Our fond thoughts go to his memory. In 1977 his Sirio satellite enabled for the first time the investigation of the 12–18 GHz frequencies for satellite communications.
  o Prof. Barry Evans (University of Surrey) and Prof. Giovanni E. Corazza (University of Bologna) who both led numerous research activities on the integration of satellites with mobile communications.
  o Dr. Walter Zoccarato and Mr. Mathieu Arnaud (Thales) who demonstrated the technical feasibility of direct satellite-to-smartphone

connectivity. Mr. Romain Bucelle (Thales) who identified the enablers for an economically viable "Direct to Device" system.
- o Mr. Laurent Combelles (Thales) and Mr. Sebastian Euler (Ericsson) for their exceptional work at ITU-R paving the way for the recognition of NTN as an IMT2020 satellite radio interface.

Acronyms

3D	Three-dimensional
3GPP	3rd Generation Partnership Project
5G	Fifth Generation
5GC	5G Core Network
5QI	5G QoS Identifier
ACK	Acknowledgment
ACLR	Adjacent Channel Leakage Ratio
ACS	Adjacent Channel Selectivity
AI	Artificial Intelligence
AM	Acknowledged Mode
AMF	Access and Mobility Management Function
AN	Access Network
AoI	Area of Interest
AR	Augmented Reality
ARP	Allocation and Retention Priority
ARQ	Automatic Repeat Request
AS	Access Stratum
BAP	Backhaul Adaptation Protocol
BCCH	Broadcast Control Channel
BCH	Broadcast Channel
BPSK	Binary Phase-Shift Keying
BWP	Bandwidth Part
CAG	Closed Access Group
CBG	Code Block Group
CBGFI	Code Block Group Flush Indicator
CBGTI	Code Block Group Transmission Information
CC	Component Carrier
CCCH	Common Control Channel
cDAI	counter Downlink Assignment Index

CE	Control Element
CFO	Carrier Frequency Offset
CHO	Conditional Handover
CM	Cubic Metric
cMTC	Critical MTC
CNR	Carrier-to-Noise Ratio
CP	Control Plane
CP-OFDM	Cyclic Prefix-Orthogonal Frequency-Division Multiplexing
CPRI	Common Public Radio Interface
CRC	Cyclic Redundancy Check
CRT	Contention Resolution Timer
CS-RNTI	Configured Scheduling Radio Network Temporary Identifier
CSI	Channel-State Information
CSI-RS	Channel-State Information Reference Signal
CU	Central Unit
DAI	Downlink Assignment Index
dB	Decibel
DC	Dual Connectivity
DCCH	Dedicated Control Channel
DCI	Downlink Control Information
DFT	Discrete Fourier Transform
DFT-S-OFDM	DFT Spread OFDM
DL	Downlink
DL-SCH	Downlink Shared Channel
DM-RS	Demodulation Reference Signals
DRB	Data Radio Bearer
DTCH	Dedicated Traffic Channel
DU	Distributed Unit
E-UTRA	Enhanced Universal Terrestrial Radio Access
E-UTRAN	Enhanced Universal Terrestrial RAN
ECEF	Earth-Centered, Earth-Fixed
ECI	Earth-Centered Inertial
eCPRI	Enhanced CPRI
eDRX	Extended Discontinuous Reception
EIRP	Effective Isotropic Radiated Power
eLTE	Enhanced LTE
eMBB	Enhanced Mobile Broadband
EMF	Electromagnetic Field
eMTC	Enhanced Machine Type Communications

EN-DC	E-UTRAN-NR Dual Connectivity
EPC	Evolved Packet Core
ESIM	Earth Station In Motion
ESOMP	Earth Station On Moving Platform
FCC	Federal Communications Commission
FDD	Frequency Division Duplex
FEC	Forward Error Correction
FFT	Fast Fourier Transform
FoV	Field of View
FR	Frequency Range
FRF	Frequency Reuse Factor
FSS	Fixed Satellite Service
FWA	Fixed Wireless Access
GCI	Geocentric Celestial Inertial
GEO	Geostationary Earth Orbit
gNB	gNodeB
GNSS	Global Navigation Satellite System
GPS	Global Positioning System
GSM	Global System for Mobile Communications
GSO	Geosynchronous Orbit
GTO	Geostationary Transfer Orbit
GW	Gateway
HAPS	High Altitude Platform System
HARQ	Hybrid Automatic Repeat Request
HD	High Definition
HEO	High Elliptical Orbit
HPA	High-Power Amplifier
HPBW	Half-Power Beam Width
I-RNTI	Inactive State Radio Network Temporary Identifier
IAA	Instantaneous Access Area
IAB	Integrated Access and Backhaul
ID	Identity
IE	Information Element
IEEE	Institute of Electrical and Electronics Engineers
iFFT	Inverse FFT
IMT	International Mobile Telecommunications
IoT	Internet of Things
IR	Incremental Redundancy
ISL	Inter-Satellite Link
ISS	International Space Station
ITU	International Telecommunications Union

ITU-R	ITU Radiocommunication Sector
kHz	kiloHertz
L1	Layer 1
L2	Layer 2
LDPC	Low-Density Parity Check
LEO	Low Earth Orbit
LHCP	Left Hand Circular Polarization
LMF	Location Management Function
LOS	Line-of-Sight
LTE	Long-Term Evolution
MAC	Medium Access Control
MBMS	Multimedia Broadcast Multicast Service
MBS	Multicast-Broadcast Services
MCC	Mobile Country Code
MCE	Multicell/Multicast Coordination Entity
MCG	Master Cell Group
MCS	Modulation and Coding Scheme
MDBV	Maximum Data Burst Volume
MEO	Medium Earth Orbit
MHz	Megahertz
MIB	Master Information Block
MIMO	Multiple Input Multiple Output
MME	Mobility Management Entity
mMTC	Massive Machine Type Communications
MN	Master Node
MNC	Mobile Network Code
MNO	Mobile Network Operator
MSG1	Message 1
MSG2	Message 2
MSGA	Message A
MSS	Mobile Satellite Service
MT	Mobile Termination
MTC	Machine Type Communications
NAK	Negative Acknowledgment or Not Acknowledged
NAS	Non-Access Stratum
NB-IoT	Narrowband Internet of Things
NCC	NTN Control Center
NCGI	NR Cell Global Identity
NCI	NR Cell Identity
NDI	New Data Indicator
NE-DC	NR-E-UTRAN Dual Connectivity

NFI	New Feedback Indicator
NG-RAN	Next-Generation Radio Access Network
NGEN-DC	NG-RAN EN-DC
NGSO	Non-Geosynchronous Orbit
NLOS	Non-Line-of-Sight
NNSF	NAS Node Selection Function
NPDCCH	NB-IoT Physical Downlink Control Channel
NPRACH	NB-IoT Physical Random Access Channel
NR	New Radio
NR-U	NR Unlicensed
Nrppa	NR Positioning Protocol a
NSA	Non-Stand-Alone
NSAG	Network Slice AS Group
NTN	Non-Terrestrial Network
NTN GW	Non-Terrestrial Networks Gateway
O-RAN	Open RAN
O&M	Operations & Maintenance
OBP	On-Board Processor
OD	Orbit Determination
OFDM	Orthogonal Frequency Division Multiplexing
OFDMA	Orthogonal Frequency Division Multiple Access
OP	Orbit Prediction
PAPR	Peak-to-Average Power Ratio
PBCH	Physical Broadcast Channel
PCC	Policy and Charging Control
PCCH	Paging Control Channel
PCH	Paging Channel
PCI	Physical Cell Identifier
PDB	Packet Delay Budget
PDCCH	Physical Downlink Control Channel
PDCP	Packet Data Convergence Protocol
PDR	Packet Detection Rule
PDSCH	Physical Downlink Shared Channel
PDU	Packet Data Unit
PHFTI	PDSCH-to-HARQ-Feedback-Timing-Indicator
PHY	Physical Layer
PLMN	Public Land Mobile Network
PLMN ID	PLMN Identifier
PO	Polar Orbit
POD	Precision Orbit Determination
PRACH	Physical Random Access Channel

PSD	Power Spectral Density
PSM	Power Saving Mode
PSS	Primary Synchronization Signal
PT-RS	Phase-Tracking Reference Signal
PUCCH	Physical Uplink Control Channel
PUSCH	Physical Uplink Shared Channel
PV	Position and Velocity
PVT	Position, Velocity, and Time
PWS	Public Warning Service
QAM	Quadrature Amplitude Modulation
QFI	QoS Flow ID
QoE	Quality of Experience
QoS	Quality of Service
QPSK	Quadrature Phase-Shift Keying
RA	Random Access
RAAN	Right Ascension of the Ascending Node
RAN	Radio Access Network
RAO	Random Access Occasion
RAR	Random Access Response
RAT	Radio Access Technology
RB	Resource Block
RE	Resource Element
RedCap	Reduced Capability
RF	Radio Frequency
RHCP	Right Hand Circular Polarization
RIM	Remote Interference Management
RLC	Radio Link Control
RMS	Root Mean Square
RMSE	Root Mean Square Error
RNTI	Radio Network Temporary Identifier
RP	Reference Point
RRC	Radio Resource Control
RRM	Radio Resource Management
RSRP	Reference Signal Received Power
RSRQ	Reference Signal Received Quality
RTD	Round Trip Delay
RTT	Round Trip Time
RU	Remote Unit
RV	Redundancy Version
S-NSSAI	Single Network Slice Selection Assistance Information
SA	Stand-Alone

SAN	Satellite Access Node
SCC	System Control Center
SCG	Secondary Cell Group
SCS	Sub-Carrier Spacing
SDAP	Service Data Adaptation Protocol
SDF	Service Data Flow
SDMA	Space Division Multiple Access
SDO	Standards-Developing Organization
SDT	Small Data Transmission
SDU	Service Data Unit
SFN	System Frame Number
SI	Study Item
SIB	System Information Block
SINR	Signal-to-Interference-Plus-Noise Ratio
SN	Secondary Node
SNO	Satellite Network Operator
SNPN	Standalone Non-Public Network
SON	Self-Organizing Networks
SPS	Semi-Persistent Scheduling
SRB	Signaling Radio Bearer
SRI	Satellite Radio Interface
SRS	Sounding Reference Signal
SS	Synchronization Signal
SSB	SS/PBCH block
SSO	Sun-Synchronous Orbit
SSP	Sub-Satellite Point
SSS	Secondary Synchronization Signal
TA	Timing Advance
TAC	Tracking Area Code
TAG	Timing Advance Group
TAI	Tracking Area Identity
TAU	Tracking Area Update
TB	Transport Block
TC-RNTI	Temporary C-RNTI
tDAI	Total Downlink Assignment Index
TDD	Time-Division Duplexing
TDMA	Time-Division Multiple Access
Te	Transmission Error
TM	Transparent Mode
TM/TC	Telemetry/Telecommand
TN	Terrestrial Network

TNL	Transport Network Layer
TP	Transmission Point
TR	Technical Report
TRP	Transmission and Reception Point
TRS	Tracking Reference Signal
TS	Technical Specification
TSG	Technical Specification Group
TT&C	Telemetry Tracking and Control
TTFF	Time to First Fix
TTI	Time Transmission Interval
UAS	Unmanned Aerial System
UCI	Uplink Control Information
UE	User Equipment
UL	Uplink
UL-SCH	Uplink-Shared Channel
ULI	User Location Information
UM	Unacknowledged Mode
UMTS	Universal Mobile Telecommunications System
UP	User Plane
UPF	User Plane Function
URLLC	Ultra-Reliable, Low-Latency Communications
UTSRP	Uplink Time Synchronization Reference Point
V2X	Vehicle-to-Everything
vLEO	Very Low Earth Orbit
VMR	Vehicle Mounted Relay
VR	Virtual Reality
VSAT	Very Small Aperture Terminal
WG	Working Group

1

Introduction

Satellite Communications (SatCom) can be defined as the use of artificial satellites to establish communication links between various points on the Earth's surface, i.e., a telecommunication system encompassing at least one communication satellite. Since the launch of the first artificial satellite (1957, Sputnik 1, Soviet Union), SatCom gained ever-increasing attention, and various artificial satellites have been launched in the following years to provide manyfold services, including communications, navigation, and Earth observation, among others. Figure 1.1 shows the number of launched satellites between 1974 and 2022 per service category; it can be noticed that the number of commercial platforms has always been larger compared to other service categories, and it is still increasing. In some cases, these platforms are multi-mission and also allow to provide government or military applications.

It is also interesting to notice the trends in terms of target orbit. Figure 1.2 shows the number of launched satellites to Geostationary Earth Orbit (GEO), Low Earth Orbit (LEO), Medium Earth Orbit (MEO), and Highly Elliptical Orbit (HEO) since 1974. Initially, taking into account the massive costs involved in both manufacturing and launches, the satellite providers aimed at deploying large platforms at high altitudes, to cover a portion of the Earth as large as possible with as few satellites as possible. In the past years, thanks to the significant advancements in the space industry in terms of on-board capabilities and cost reduction (in particular, thanks to the rise of reusable first-stage rockets), we have witnessed a surge in the research and development of LEO and Very Low Earth Orbit (VLEO) systems, followed by a new space race to mega-constellations (e.g. SpaceX Starlink, OneWeb, and Amazon Kuiper, to name a few).

In parallel to this trend, the past years have also seen significant advancements related to Air-to-Ground (ATG) systems, drones, and High-Altitude Platform Stations (HAPS), initially to support the connectivity to on-ground

5G Non-Terrestrial Networks: Technologies, Standards, and System Design, First Edition.
Alessandro Vanelli-Coralli, Nicolas Chuberre, Gino Masini, Alessandro Guidotti, and Mohamed El Jaafari.
© 2024 The Institute of Electrical and Electronics Engineers, Inc. Published 2024 by John Wiley & Sons, Inc.

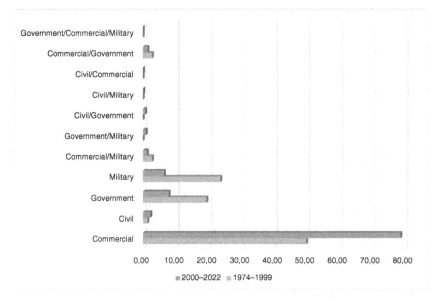

Figure 1.1 Percentage of launched satellites per service type between 1974 and 2022. Source: Adapted from https://www.ucsusa.org/resources/satellite-database.

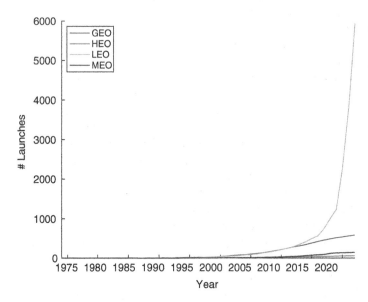

Figure 1.2 Number of launched satellites per orbit. Source: Adapted from https://www.ucsusa.org/resources/satellite-database.

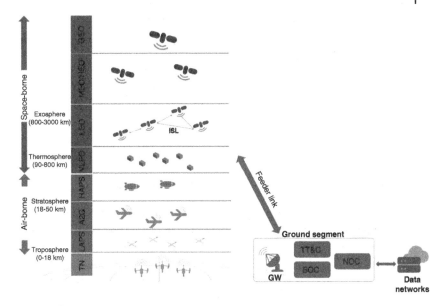

Figure 1.3 From SatCom to NTN.

users in critical scenarios, but now also for many other applications. The combination of the New Space era with the exploitation of low-altitude nodes led to a broadening in the family of communications from the sky: Non-Terrestrial Networks (NTN), which include the use of airborne and spaceborne nodes to establish communication links between various points on the Earth's surface. As shown in Figure 1.3, we now have a continuum of layers/orbits in which NTN nodes can be deployed, allowing terrestrial networks (TNs) to really touch the sky.

1.1 What is 5G NTN?

Throughout history, legacy SatCom systems have been designed and developed relying on industry-driven technical specifications leading to proprietary architectures, protocol stacks, and radio access solutions, which made the interoperability of the satellite access network between different vendors not granted. This approach of the SatCom Industry leads to a fragmented market with vendor-locked solutions, and limited interworking with mobile systems. With the recent publication of 3GPP technical specifications on the introduction of NTN, a global standard for satellite systems has been newly defined aiming at supporting any orbit, any frequency band, and any device. Such specifications open the door for the seamless integration

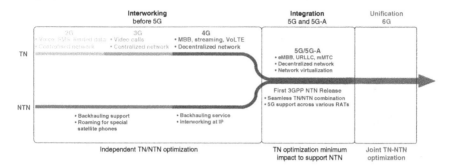

Figure 1.4 Toward a network unifying NTN–TN network components.

of airborne and spaceborne network components in the 5G system (5GS) and beyond (including 5G-Advanced and the upcoming 6G), delivering the promise of a ubiquitous mobile system that can support new use cases.

As depicted in Figure 1.4, before 3GPP Release-17, 3GPP networks were natively designed and optimized only for terrestrial-based cellular networks. On the other hand, Satellite networks, as previously mentioned, were based on proprietary technologies. Thereby, only limited interworking between SatCom-based networks and 3GPP mobile network components was possible, mainly including backhaul services. The massive 3GPP work on NTN, and the resulting integration of the satellite technology in 3GPP specifications starting from 3GPP Release-17, opened a brand new frontier in 3GPP cellular systems and ushered in new paradigms for connected society, thereby delivering the promise of a ubiquitous end-to-end ecosystem that can support a myriad of new use cases. Here, "integration" means that the space-/air-borne and terrestrial components of the network are able to seamlessly work together to provide coverage continuity to the end users. As the 5G design has originally been optimized for the TN component, a great care has been taken to minimize impacts on User Equipment (UE), 5G Radio Access Network (RAN), and 5G Core (5GC) network level, while supporting the largest range of NTN deployment scenarios.

In order to support such hybrid terrestrial-satellite[1] systems enabling New Radio (NR) and Internet of Things (IoT) services through satellites, the 3GPP work started with Study Items (SIs) on NTN in Releases 15 and 16, as shown in Figure 1.5; the necessary features for the support of the NTN component have been then specified as part of the 3GPP Release-17. The Release-17 normative works on NTN and satellites in 3GPP Technical Specification Group (TSG) RAN and TSG Service and system Aspects (SA) have been completed

---

1 Throughout this book, if not otherwise specified, "satellite" elements are to be intended in the 3GPP NTN meaning, i.e., spaceborne and airborne components.

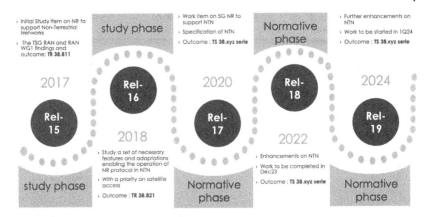

Figure 1.5 3GPP Works on NTN up to Rel-18. Source: El Jaafari et al. [1].

in June 2022 and the related ASN.1 freeze in September 2022 [2]. As previously mentioned, particular attention was given to the minimization of the impacts at UE, NG-RAN, and 5GC levels to support NTN by re-using as much as possible the existing terrestrial-optimized 5G specifications. This first 3GPP NTN standard provides the specifications for 3GPP satellite access networks based on both 5G NR protocols and 4G Narrowband IoT (NB-IoT) and enhanced Machine Type Communications (eMTC) radio protocols; in both cases, the system is operating in Frequency Range 1 (FR1)[2] bands. In Release-17, the NR-based satellite access is designed to serve handheld devices to provide enhanced Mobile Broadband (eMBB) services, while the NB-IoT/eMTC-based satellite access aims at providing MTC services to IoT devices for applications in agriculture, transport, logistics, and security markets. To support new scenarios and deployments above 10 GHz, as well as to introduce several enhancements for NR-NTN and IoT-NTN, a normative work is currently being carried out as part of Release-18.

The support of IoT-NTN is largely aligned with that of NR-NTN in 5GS. It is worthwhile mentioning that, since access networks based on Unmanned Aerial System (UAS), including HAPS and drones, could be considered as a special case of NTN access with lower latencies and Doppler values and variation rates, the main focus is on satellite-based NTN only.

In 3GPP, NTN refers to a network providing non-terrestrial access to UEs by means of an NTN payload embarked on an air-borne or space-borne NTN vehicle and an NTN gateway. A space-borne vehicle can embark a bent-pipe or a regenerative payload telecommunication transmitter, and it

---

2 What we usually call sub 6-GHz; referred to as FR1. The corresponding frequency range is from 410 to 7125 MHz.

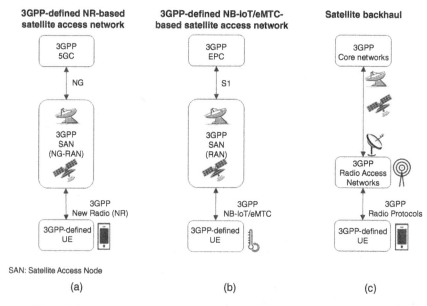

3GPP-defined NR-based satellite access network

3GPP-defined NB-IoT/eMTC-based satellite access network

Satellite backhaul

SAN: Satellite Access Node

(a)   (b)   (c)

Figure 1.6 3GPP defined satellite network solutions for 5G. Source: El Jaafari et al. [1].

can be placed at LEO, MEO, or GEO orbits; an air-borne vehicle is HAPS encompassing UAS, including Lighter than Air UAS (LTA) and Heavier than Air UAS (HTA), all operating in altitudes typically between 8 and 50 km, i.e., quasi-stationary.

As shown in Figure 1.6, the following satellite network solutions have been integrated within 5GS starting from 3GPP Release-17:

(a) 3GPP-defined NR-based satellite access network: NG-RAN based on satellite access nodes (SANs), connected to a 5GC, providing eMBB via satellite (eMBB-s) and High Reliability Communication via satellite (HRC-s) services to 3GPP-defined UEs. It supports the 3GPP NR access technology and it may also provide connectivity to Integrated Access and Backhaul (IAB) nodes.

(b) 3GPP-defined LTE-based satellite access network: Evolved Universal Terrestrial Radio Access (E-UTRA) RAN-based on SANs, connected to an Evolved Packet Core network (EPC), providing mMTC via satellite (mMTC-s) services to 3GPP UEs. It supports the 4G NB-IoT/eMTC access technology.

(c) Satellite backhaul: A transport network over satellite that provides connectivity between the 5GC and the gNB, which can be based on 3GPP or non-3GPP-defined radio protocols.

1.2 Use Cases for 5G NTN

The emergence of hybrid terrestrial-satellite systems is the result of a joint effort between stakeholders of both mobile and satellite industries, and it is paving the way to new business opportunities. 3GPP Technical Report (TR) 38.811 [3], reports the identified use cases for the provision of services when considering the integration of the NTN access component in the 5GS; a more detailed description of each use case can be found in 3GPP TR 22.822 [4]. The identified use cases benefit from the wide service coverage capabilities and the reduced vulnerability provided by the space-/air-borne nodes of the NTN component to physical attacks and natural disasters. In general, they can be categorized into the following three macro-usage scenarios:

- Service ubiquity and global connectivity: This category targets the reduction of the digital divide by providing direct access connectivity to handheld terminals, households, and IoT devices. This scenario will complement the TN in under-served or un-served geographical areas. Some uses cases that fall into this category include direct access to smartphone and home access in rural or isolated areas, Public Safety and Public Protection and Disaster Relief (PPDR) in remote areas or areas that experienced a disaster (for instance, earthquakes, floods, or terrorist attacks) leading to a partial or total destruction of the terrestrial infrastructure, IoT for agriculture, critical infrastructures metering and control–pipelines, and asset tracking/tracing.
- Service continuity and resiliency: This scenario supports connectivity in situations or areas where 5G services cannot be offered by TNs alone. A combination of terrestrial and non-terrestrial elements provides service continuity, higher availability, and reliability. Some potential use cases include connectivity to moving terrestrial platforms (e.g. trains, cars, and trucks), maritime platforms, and airborne platforms (e.g. aircraft).
- Service scalability: Thanks to the inherently large coverage footprint of satellites, NTN is very efficient in providing multicast and broadcast services to the users. Some use cases that might take advantage of this feature include, but are not limited to: Public Safety and mission critical, massive software delivery over NTN, group communications, media and entertainment (e.g., live broadcasts), and ad-hoc broadcast/multicast streams.

An overview of both interworking and integration of TNs and NTN is shown in Figure 1.7, including interworking and integration through direct and indirect connectivity.

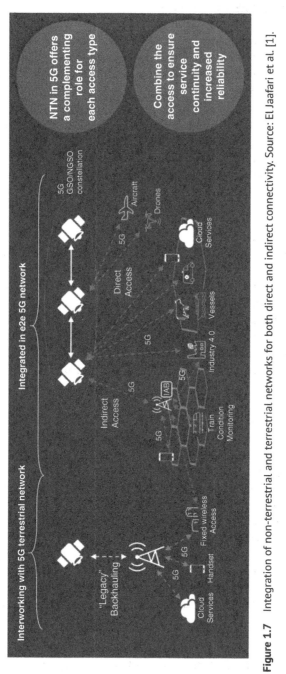

Figure 1.7 Integration of non-terrestrial and terrestrial networks for both direct and indirect connectivity. Source: El Jaafari et al. [1].

1.3 ITU-R Vision and Requirements on the Satellite Component of IMT-2020

The International Telecommunication Union Radiocommunication (ITU-R) sector follows a well-established procedure to define requirements, request submissions by external organizations (e.g. 3GPP), and evaluate radio interface technologies for the definition of International Mobile Telecommunications-2020 (IMT-2020). Such principles and procedures are reported in ITU-R Resolution 65 [5], and they are applicable to both the terrestrial and NTN components of IMT-2020.

Within this framework, ITU-R has developed Report M.2514 [6], in which the vision and the requirements of the satellite radio interfaces for IMT-2020 are described. This report represents the baseline document for the submission of the IMT-2020 satellite component radio interface technologies to ITU-R.[3]

1.3.1 Satellite Component of IMT-2020: Usage Scenarios

Recommendation ITU-R M.2083 on the IMT-2020 vision [7], identifies three usage scenarios for IMT-2020 and beyond: eMBB, mMTC, and Ultra Reliable and Low Latency Communications (URLLC). The satellite component of IMT-2020 is expected to provide eMBB-s and mMTC-s services, i.e., NTN-based variants of the eMBB and mMTC defined in Recommendation M.2083. The suffix -s (via satellite) of the naming captures the satellite specificity in terms of expected performance. Moreover, the satellite component of IMT 2020 will not cover URLLC scenarios; however, it will cover HRC-s scenarios. The diagram in Figure 1.8 depicts the usage scenarios for the satellite component of IMT-2022, together with some of the corresponding use cases.

1.3.2 Requirements for the Satellite Radio Interface(s) of IMT-2020

Satellite systems are characterized, among others, by wide coverage, resiliency, and multicast/broadcast capabilities; as such, they will provide scalable and efficient network solutions for 5G and beyond communications. The satellite component of IMT-2020 will enhance the benefits of the

---

3 The detailed process for the submission of radio interface technologies is detailed at the following link: https://www.itu.int/en/ITU-R/study-groups/rsg4/Pages/imt-2020-sat-submission-eval.aspx.

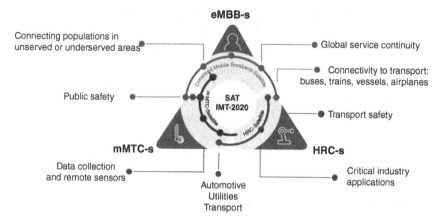

Figure 1.8 Satellite component of IMT-2020 usage scenarios. Source: El Jaafari et al. [1].

IMT-2020 use cases and their corresponding service requirements. In this framework, both Geosynchronous Orbit (GSO) and Non-Geosynchronous Orbit (NGSO) mobile satellite systems will have a role to play.

In terms of technical performance parameters, those adopted for the IMT-2020 satellite component are similar to those initially defined by ITU for TNs; this allows to have consistency and alignment between the two network components. Nonetheless, for the satellite component, the target performance associated with these metrics has been adapted to take into account the peculiarities of satellite systems. The diagram in Figure 1.9

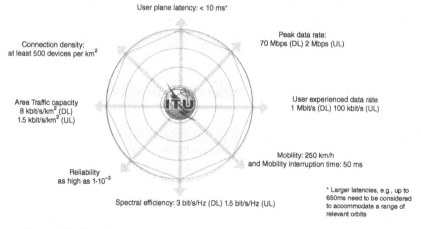

Figure 1.9 Requirements for the satellite radio interface(s) of IMT-2020. Source: El Jaafari et al. [1].

summarizes the main performance requirements that are expected from the satellite component of IMT-2020, defined assuming channel bandwidths up to 30 MHz.

1.4 NTN Roadmap in 3GPP

For the first time in 3GPP history, in March 2017, during the RAN#75 meeting in Dubrovnik, satellite industries and stakeholders were given the unique opportunity to be involved in the standardization process of a new standard integrating terrestrial and NTNs. Since then, satellite-based access has been included in the 3GPP roadmap, marking a radical paradigm shift with respect to how mobile network standards are conceived.

In Release-16, several SIs have been conducted in different Working Groups (WG). These are listed hereafter:

- SA1: Study on using satellite access in 5GS aimed at identifying the use cases for the provision of services in scenarios integrating the 5G satellite-based access components in the 5GS. The system and service and performance requirements for such satellite-based 5G access have also been defined in SA1;
- SA2: Study on architecture aspects for the introduction of the satellite component in the 5G access network;
- SA5: Study on management and orchestration aspects for the satellite components integrated into the 5GS;
- RAN: Based on the outcomes of the study phase in Release-15, the RAN WGs performed a study on the challenges and related solutions for the support of NTN in the NR standard.

In Release-17, based on the outcomes of the extensive study phase in Rel.15 and 16, 3GPP conducted the normative work on both NR-NTN and IoT-NTN:

- SA2: Work Item on the integration of the non-terrestrial component in the 5G architecture;
- Core network and Terminals 1 (CT1): Activities related to the study and corresponding specifications of the solutions for the integration of satellites in the 5G architecture and the support of Public Land Mobile Network (PLMN) selection for satellite access;
- RAN: Work Item on the 5G NR support of the NTN component; some initial studies were performed in Release-17 for IoT-NTN as well, followed by a normative work, to specify both NB-IoT and eMTC support for NTN.

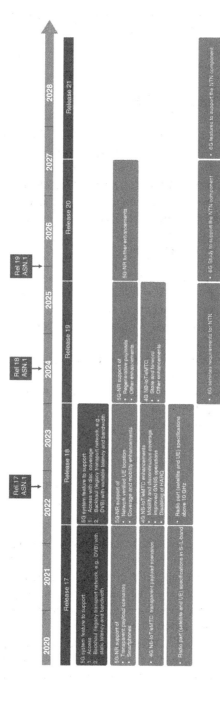

Figure 1.10 Satellite-related study and work items in 3GPP. Source: Guidotti et al. [8].

As previously highlighted, the first set of technical specifications reporting the required features and adaptations for the support of NTN in 5G was completed in June 2022 within Release-17, while the ASN.1 freeze was concluded in September 2022.

At the time of writing this book, 3GPP activities are now devoted to Release-18, which started in May 2022. With respect to the support of NTN, this Release includes Work Items on further enhancements for both NR and NB-IoT/eMTC over NTN. Based on the original RAN Release-18 timeline, March 2024 is the target date for ASN.1 freeze. The 3GPP roadmap is shown in Figure 1.10, together with a list of the main NTN objectives per Release.

1.5   3GPP Requirements for 5G via Satellite

Building on the outcomes of TSG SA1 studies collected in TR 22.822, 3GPP defined a list of target requirements and Key Performance Indicators (KPIs) to support the identified use cases for satellite-based access in TS 22.261 [9]. The most relevant ones include the following:

- A 5GS providing service with satellite access shall be able to support LEO, MEO, and GEO-based satellite access with up to 35, 95, and 285 ms end-to-end latency, respectively.
- To optimize the Quality of Experience (QoE) for the UE, a 5GS shall support negotiation on quality of service (QoS) taking into account the latency penalty.
- The 5GS with satellite access shall support communication service availabilities of at least 99.99%.
- The 5GS with satellite access shall support high uplink and downlink data rates for 5G satellite UEs.

The performance requirements for satellite access along with the defined target values are available in table 7.4.2-1 in [9].

1.6   Technical Challenges

To integrate satellite-based access in the 5GS, the 3GPP standardization activities clearly took into account the specific characteristics and features of a wide range of satellite network deployments. These create significantly different and new challenges compared to those in terrestrial mobile networks. In particular, the issues related to long propagation delays, large Doppler effects, and the generation of large moving cells on-ground had to be tackled. Table 1.1 reports a list of the technical issues posed by satellite

Table 1.1 LEO Network specific characteristics creating and related technical issues.

LEO network characteristics	Technical issues	Comments
Motion of the satellites	Dynamic cell pattern	Steerable beams are generated by the satellite to provide quasi-fixed beam footprints on-ground or satellite-fixed beams are generated by the satellite and, hence, the beam footprint pattern moves on the ground with the satellite motion
	Delay variation	The propagation delay varies proportionally with the variation of the distance between the satellite and the UE. The maximum delay variation as seen by the UE can be up to ± 47.6 μs/s[a)]
	Doppler	The satellite motion causes a Doppler shift, as well as a Doppler variation, proportional to the radial satellite velocity as seen by the UE.
Altitude of satellites	Latency	The propagation delay directly depends on the satellite altitude, the minimum elevation angle of the satellite as perceived by the UE, and whether the radio link with the UE is terminated at on-board or on-ground (at the gateway side). The maximum distance between the satellite and the UE at minimum elevation angle can be up to 1932 km (600 km altitude), 3131 km (1200 km altitude) and 40 581 km (for GEO based NTN). Thereby, for transparent payload, the maximum Round Trip Delay is 25.77 ms for LEO600km, 41.77 ms for LEO1200km, and 541.46 ms for GEO.
Beam size	Differential delay	The difference in delay experienced by a UE at the beam edge versus at beam center is proportional to the beam size. The maximum differential delay could be equal to 3.12 and 3.18 ms for respectively LEO600km and LEO1200km and equal to 10.3 ms for GEO-based NTN.
Propagation channel	Frequency selectiveness impairments	Outdoor usage conditions are always assumed in satellite networks with a quasi-Line-of-Sight (LoS), where most attenuation comes from atmospheric conditions and possible foliage creating low shadowing. It thus results in a flat channel characteristic.

Table 1.1 (Continued)

LEO networ characteristics	Technical issues	Comments
	Delay spread impairments	Unlike cellular networks, this has a negligible effect on satellite networks. The maximum delay spread ranges from 10 ns in Ka band to 100 ns in S band [3].
Duplex scheme	Regulatory constraints	Most spectrum allocated to satellite services feature paired bands for Space-to-Earth and Earth-to-Space, respectively, suited for Frequency Division Duplexing (FDD) mode. This is due to the fact that, unless a low-altitude platform with on-board base station can be considered, a significant guard time between uplink and downlink would have to be provisioned in the case of Time Division Duplexing (TDD).

a) Maximum Round Trip Delay variation as seen by the UE is up to ±93.0 μs/s (Worst case) in case of a LEO transparent payload and up to ±47.6 μs/s for a LEO regenerative payload.

radio links compared to cellular networks, which were initially captured and discussed in TR 38.811 [3], and TR 38.821 [10].

1.7 Satellite RAN Architecture

Since Release-17, PLMN core networks can be connected to a space-/air-born RAN, which can be shared between two or more core networks. Figure 1.11 shows the NG-RAN (a) and EUTRAN (b) that support the implementation of NTN with transparent payloads. The NG-RAN in Figure 1.11(a) encompasses a set of SAN connected to the 5GC through the NG Air Interface. A SAN provides the NR User Plane (UP) and Control Plane (CP) terminations towards an NTN-enabled UE, which can access the NTN services through the payload via the service link, and it includes a transparent NTN payload on-board the NTN platform, a gateway, and gNB functions. It shall be mentioned that a single gNB might serve multiple NTN payloads and a single NTN payload might be served by multiple gNBs on-ground, depending on the specific NTN system design. The NTN payload transparently forwards the radio protocols that are received from the UE (connected via the service link) to the NTN gateway (via the feeder link) and *vice versa*. Since a transparent payload is considered, the only available operations on-board include filtering, frequency conversion,

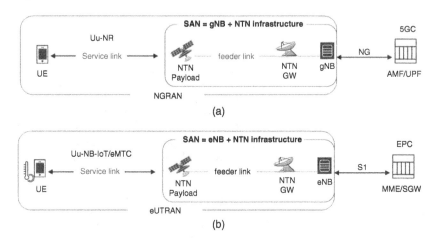

Figure 1.11 NTN based RAN architecture with transparent satellite. Source: El Jaafari et al. [1]. (a) NTN NG-RAN architecture and (b) IoT NTN EUTRAN architecture.

and power amplification, before the re-transmission on the service/feeder link is performed.

The following types of cell coverage are supported:

- **Earth-fixed coverage**, provided by GSO satellites generating beam(s) that continuously cover the same geographical areas for the entire lifetime.
- **Quasi-Earth-fixed coverage**, provided by NGSO satellites creating beam(s) that cover a given geographical area for a limited period and a different one in another period, depending on their visibility. For this type of coverage, beam steering is required on-board.
- **Earth-moving coverage**, in which the beam(s) generated by the NGSO satellites sweep the Earth's surface according to the satellites' movement along their orbit, i.e., the coverage is always centered at the Sub Satellite Point (SSP).

Service and feeder link switch-over procedures are supported in the NTN standard and they typically apply in case of NGSO systems: a service link switch refers to a change of serving satellite, and a feeder link switch refers to the procedure in which the feeder link is changed from a source NTN gateway to a target NTN gateway for a given NTN payload (this is a Transport Network Layer procedure). Both hard and soft feeder link switch-over solutions are applicable to NTN.

In terms of the UE, only those equipped with GNSS (Global Navigation Satellite Systems) capabilities are supported. Both Very Small Aperture Terminal (VSAT) devices with directive antennas (including fixed and moving platform-mounted devices) and commercial handset terminals (e.g., Power Class 3) are supported in FR1 (below 7.125 GHz). As for FR2 (above 10 GHz), the support currently foresees only VSAT devices. As discussed below, to support NR-NTN access, the NTN UE shall support uplink time and frequency pre-compensation, timing relationship enhancements, and several NTN essential features, such as timer extension in Medium Access Control (MAC)/Radio Link Control (RLC)/Packet Data Convergence Protocol (PDCP) layers and Random Access Channel (RACH) adaptation to handle the long Round Trip Time (RTT), acquiring NTN specific System Information Block (SIB) and more than one Tracking Area Code (TAC) per PLMN broadcast in one cell. The NTN UE may support the NTN features in GSO or NGSO scenarios or both.

As for NB-IoT/eMTC support for NTN, the EUTRAN in Figure 1.11(b) consists of a set of SAN nodes connected to the EPC via the S1 interface. Only Bandwidth-reduced Low complexity (BL) UEs are considered, with support of UEs in enhanced coverage and NB-IoT UEs with GNSS capabilities.

1.8 NTN Spectrum

With respect to spectrum, the frequency bands supported for NTN in Release-17 are the Mobile Satellite Service (MSS) S-band n256 and L-band n255, which constitute the lower frequency bands for handheld terminals. The NTN operating bands supported in FR1-NTN[4] are given in Table 1.2. Enhancements to support new scenarios for NR-NTN deployments in FR2-NTN[5] are being specified in Release-18 with the harmonized ITU-R Ka-band serving as a reference. The operating bands supported for NTN in FR2-NTN are given in Table 1.3. In all these bands Frequency Division Duplex mode is considered.

During the WRC-23 held from November to December 2023 in Dubai, ITU-R, supported by worldwide delegates, defined all aspects related to the

---

4 Corresponding frequency range: 410–7125 MHz. NTN bands within this frequency range are regarded as a FR1 band when references from other specifications.
5 FR2-NTN (17 300–30 000 MHz) is the NTN frequency range above 10 GHz defined so far. NTN bands within this frequency range are regarded as a FR2 band when references from other specifications.

Table 1.2 NTN operating bands in FR1-NTN.

NTN operating band	Uplink (UL) operating band SAN receive/UE transmit $F_{UL,low} - F_{UL,high}$	Downlink (DL) operating band SAN transmit/UE receive $F_{DL,low} - F_{DL,high}$	Duplex mode
n256	1980–2010 MHz	2170–2200 MHz	FDD
n255	1626.5–1660.5 MHz	1525–1559 MHz	FDD
n254	1610–1626.5 MHz	2483.5–2500 MHz	FDD

Table 1.3 NTN operating bands in FR2-NTN (under definition).

NTN operating band	Uplink (UL) operating band SAN receive/UE transmit $F_{UL,low} - F_{UL,high}$	Downlink (DL) operating band SAN transmit/UE receive $F_{DL,low} - F_{DL,high}$	Duplex mode
n512[a]	27.5–30.0 GHz	17.3–20.2 GHz	FDD
n511[b]	28.35–30.0 GHz	17.3–20.2 GHz	FDD
n510[c]	27.5–28.35 GHz	17.3–20.2 GHz	FDD

a) This band is applicable in the countries subject to CEPT ECC Decision(05)01 and ECC Decision (13)01.
b) This band is applicable in the USA subject to FCC 47 CFR part 25.
c) This band is applicable for Earth Station operations in the USA subject to FCC 47 CFR part 25. FCC rules currently do not include ESIM operations in this band (47 CFR 25.202).

use of spectrum for the next years and drafted the Agenda Items for further study to be discussed during WRC-27 and WRC-31.

The detailed outcomes of WRC-23 are available in the ITU-R WRC-23 [11], booklet and in the CEPT weekly reports [12]. In this framework, the following topics related to NTN have been discussed: frequency allocations for HAPS, harmonisation of Ku-band spectrum, spectrum and interference management in Ka-band, NGSO-NGSO and GSO-NGSO ISL operations, new allocations to fixed-satellite services in Ka and Ku bands.

Among the topics identified for WRC-27 and WRC-31, the following shall be mentioned: Q/V bands allocations on the feeder link, new Ka-band allocations for the fixed-satellite service, and the identification of potential new allocations for direct connectivity, mobile-satellite services, and ISLs. Table 1.4 reports the FRs currently envisaged for SatCom above 10 GHz.

With respect to HAPS, in general, they can use the spectrum allocated to terrestrial systems, with the potential modifications based on the WRC-23 outcomes mentioned above and the studies that will be performed for WRC-27 and WRC-31.

Table 1.4 ITU-R frequency ranges above 10 GHz for Satellite Communications.

Operating band	Uplink (Earth-to-Space)	Downlink (Space-to-Earth)
Ku-band	12.75–13.25 GHz 13.75–14.5 GHz	10.7–12.75 GHz
Ka-band (GEO)	27.0–30.0 GHz	17.3–20.2 GHz
Ka-band (NGSO)	27.0–29.1 GHz 29.5–30.0 GHz	17.7–20.2 GHz
Q/V-band	42.5–43.5 GHz 47.2–50.2 GHz 50.4–52.4 GHz	37.5–42.5 GHz 47.5–47.9 GHz 48.2–48.54 GHz 49.44–50.2 GHz

1.9 3GPP Work on NTN in Release-15 and Release-16

The activities related to NTN in 3GPP Release-15 and 16 (study phase) are summarized in Table 1.5 and reported below.

Table 1.5 List of satellite-related activities in 3GPP in the different TSG in Release-15 and Release-16.

3GPP Release	RAN	SA
Rel.15	• Initial study item (SI) focusing on the definition of the NTN propagation channel model and on the identification of the specifications of NTN compared to cellular networks. • Outcomes reported in TR 38.811.	• SA1: alignment of service requirements for satellite access and fixed broadband access. • Outcomes captured in TS 22.261 and SP-180997
Rel.16	• Second study characterized the NTN specifications (e.g. long propagation delays, large Doppler effects, and moving cells) and the related solutions to address them.	• SA WG1: definition of the satellite-related service requirements. Outcomes provided in 3GPP TS 22.261. • SA WG1: study on using satellite access in 5G (identification of the use cases for the satellite network component of 5G). Findings captured in 3GPP TR 22.822

Table 1.5   (Continued)

3GPP Release	RAN	SA
	• Outcomes reported in 3GPP TR 38.821	• SA WG2: identification of the issues and potential solutions associated with the architecture aspects of using satellite access in 5G. Outcomes reported in 3GPP TR 23.737. • SA WG5: identification of the issues and potential solutions associated with the management and orchestration aspects of integrated satellite components in a 5G network. Findings reported in 3GPP TR 28.808.

In the context of SA1, TS 22.261 [9], reports the system and service, and performance requirements for 5G satellite access defined within Release-16, based on an alignment of the service requirements discussed in Release-15 and captured in SP-180997 [13]. In addition, a study related to the exploitation of satellite access in 5GS was conducted in SA1 aimed at the identification of the use cases for the provision of services via NTN integration and at addressing the requirements on set-up, configuration, and maintenance of the UE features when it exploits NTN components combined with other 5GS components from the 5GS, as well as regulatory requirements when moving to/from the satellite from/to the TNs. All TSG SA1 outcomes and discussions are collected in 3GPP TR 22.822 [4].

At 5GS level, a SI covering architecture aspects for satellite-based access in 5G was conducted in SA2 within Release-16. During this study, taking as an input the use cases from TR 22.822, a set of impact areas was identified and proper solutions to address the satellite-specific key issues and solve RAN and CN inter-related issues were defined. Such SA2 findings and outcomes are reported in 3GPP TR 23.737 [14].

Release-16 also included a study related to management and orchestration aspects for the integrated satellite components, led by SA5. The scope of this study was to identify the most relevant key issues associated with business roles, service and network management, and orchestration of 5G networks, including the NTN component(s); the NTN component could be included as NG-RAN, for non-3GPP access, or transport. Moreover, the SA5 study also identified the associated solutions. The outcomes of this analysis are collected in 3GPP TR 28.808 [15].

Focusing now on RAN activities, an initial SI on the support of NTN in NR was conducted in the context of Release-15. This study was aimed at defining the NTN deployment scenarios and the related system parameters, including architecture and orbit, as well as to adapt the 3GPP channel models to the NTN channel characteristics (including propagation conditions and mobility). This SI also identified the most critical impact areas of the NTN characteristics on the NR Air Interface that required further investigations. The outcomes and findings of this study are captured in TR 38.811 [3].

Based on the Release-15 outcomes reported in TR 38.811, within Release-16, the RAN WGs also performed a study to identify the solutions to be implemented in NR to support NTN. The objective of this assessment was the study of the required features and adaptations that would enable NR protocols on NTN, with satellite access being prioritized. More specifically, the objectives were: (i) the consolidation of the potential impacts at physical layer level and the corresponding solutions; (ii) the performance assessment of NR in a set of selected scenarios, which included LEO and GEO satellite access, through both link (radio link) and system (cell) level simulations; and (iii) the identification of issues and solutions on NR related to layers 2 and 3. The outcomes of these extensive analyses were reported in 3GPP TR 38.821 [10].

Table 1.5 reports a list of all satellite-related activities within 3GPP Release-15 and 16.

1.10   3GPP work on NTN in Release-17 and Release-18

Building on the massive work performed in Releases 15 and 16 (study phase), the normative phase in Release-17 provided several enhancements at the system architecture level to address: (i) mobility management (MM) aspects for large coverage areas; (ii) MM aspects for moving coverage areas; (iii) large delays over the NTN channel; (iv) QoS with satellite access (which can be via NR or NB-IoT/eMTC as reported in Figure 1.6) and with satellite backhaul; and (v) regulatory services with super-national satellite ground stations. In this framework, the following topics were addressed:

- With respect to 5GS MM procedures, which include handover management for the radio bearer between the nodes and paging management (i.e. the management of the reachability of the UE for downlink services), adaptations have been introduced to take into account the beam footprint size and the presence of both fixed or moving cell configurations.

- Both Tracking Areas and Cell Identities (cell IDs) have been agreed to refer to specific geographical areas, in order to let the 5GS exploit these identifiers as an indication of the UEs' location.
- With moving cells, in order to avoid TAC fluctuations, it was agreed that the RAN shall broadcast the list of TACs within the cell. Such list corresponds to the Tracking Areas that have been defined on the Earth's surface based on the network planning related to the area currently illuminated by the radio cell.
- New 5G QoS classes (5QI) have been introduced to support the increased latency for satellite access and backhaul.
- New Radio Access Technologies (RAT) types have been introduced that are capable of discerning between the different types of satellite access, i.e., LEO, MEO, GEO, or others.

In the framework of Release-17, SA2 also defined the solutions at 5G architecture level for the support of satellite backhauling, e.g., the connection between gNBs and 5GC via a single GEO or a single NGSO satellite as shown in Figure 1.6(c). In order to limit as much as possible the impact on the NR architecture, only backhauling with constant delay was considered; as a consequence, it is possible to mask any latency modification over the service or feeder link through the exploitation of the knowledge of the satellite ephemeris, which allows to pre-compute the amount of delay that is needed to maintain the overall delay constant. To support satellite backhauling, the following new features were introduced in TS 23.501 [16], on the QoS for satellite backhauling: (i) reporting the satellite backhaul category to the 5G Core; and (ii) new 5QI defined specifically for satellite access and backhaul, as previously mentioned.

In Release-17, within CT1 the objective has been that of studying and defining the specifications to support the PLMN selection for satellite access. The outcomes of these studies are provided in 3GPP TR 24.821 [17], which include:

- New NG-RAN satellite RAT type in the Universal Subscriber Identity Module (USIM).
- The extension of Non-Access Stratum (NAS) supervision timers over satellite access for GEO/MEO access types, with LEO exploiting legacy timers.
- The modification of the higher priority PLMN selection procedure to include shared Mobile Country Code (MCC) 9xx.

- The introduction of new minimum periodic search timers for higher priority PLMN search through satellite access, when the PLMN is using shared MCC.
- New triggers for the selection of the PLMN upon the transition in or out of international areas, which is based on UE implementation.
- The introduction of a new list of forbidden PLMNs, not allowed to operate at the UE's location, and its related handling.
- New 5G MM cause value#78 and its related handling (this is related to the above feature).
- The support of multiple TACs for the same PLMN broadcast in the radio cell, which includes the related logic for the determination of the current Tracking Area Identity (TAI) and the impact on the Mobile Equipment (ME)-USIM procedures.

With respect to RAN, the Work Item on NR-NTN was approved at RAN meeting #86 in December 2019 [18], aiming at specifying the enhancements to support the NTN component by addressing the technical challenges reported in Table 1.1. A large number of enhancements were introduced to support NTN radio access solutions in LEO and GEO systems, with implicit compatibility to support HAPS and ATG scenarios. The main assumptions for NR-NTN in Release-17 are the following:

- The NTN node is equipped with a transparent payload.
- The targeted UEs are handheld devices (e.g., Power Class 3) operating in FR1, in particular S or L band.
- The UE terminal shall be equipped with a GNSS receiver, thus being able to estimate its location.
- Only FDD mode is considered.

At physical layer, Release-17 specified a set of enhancements to meet the previously mentioned challenges introduced by the satellite radio links in terms of large RTT, timing drift, and Doppler shift/variation:

- In order to acquire uplink time and frequency synchronization, the NTN UE shall support their pre-compensation. In particular:
  o The UE shall calculate a UE specific Timing Advance (TA), which corresponds to the RTT on the service link, based on the location acquired through GNSS and the serving satellite ephemeris;
  o The UE shall also compute the common TA, i.e., any common delay such as the RTT on the feeder link, according to the parameters provided by the network (the UE considers a null common TA if such parameters are not provided);

- o The UE shall perform a pre-compensation of the computed TA on its uplink transmissions;
- o For the TA update when the UE is in connected mode (RRC_ CONNECTED), the combination of open control loops (autonomous estimate of the TA) and closed control loops (received TA commands) is supported;
- o Frequency pre-compensation is supported to cope with the Doppler shift experienced on the service link.
- – The 5G network broadcasts assistance information (i.e., the ephemeris and the common TA parameters) within the serving NTN cell. The UE shall acquire a valid GNSS position as well as the assistance information before and during connection to an NTN cell.
- – While the pre-compensation of the instantaneous Doppler shift experienced on the service link is to be performed by the UE for the uplink, the management of Doppler shift experienced over the feeder link is left to the network implementation.
- – Further, to cope with the propagation delay on NTN link, the UE shall support several enhancements on the timing relationships:
  - o Many timing relationships, such as the Physical Uplink Shared Channel (PUSCH) transmission timing relationship, are enhanced by introducing a scheduling offset denoted as K_{offset};
  - o MAC CE timing relationships are also enhanced with an offset (i.e. k_mac) when downlink and uplink frame timing are not aligned at gNB.
- – To avoid Hybrid Automatic Repeat request (HARQ) stalling in NTN, which might be caused by the large RTT, a hybrid approach is adopted:
  - o The HARQ feedback can be dynamically disabled in the presence of ARQ re-transmissions at RLC layer (e.g. GSO systems);
  - o The number of HARQ processes for re-transmissions at MAC layer can be increased to up to 32 (e.g. NGSO systems).

Also, higher layers have been adjusted and enhanced on both the UP and CP to adapt them to the NTN characteristics:

- – UP enhancements:
  - o For the Random Access (RA) procedure, the UE shall support the estimation of the RTT to the gNB and delay the start of the RA Response (RAR) window by this amount;
  - o The UE reporting information related to TA pre-compensation shall be supported as per TS 38.321 [19];
  - o To limit the detrimental impact of HARQ stalling, which can significantly reduce the peak throughput, two features have been introduced,

as previously mentioned: (i) disable/enable the HARQ feedback per process; or (ii) extend the maximum number of HARQ processes to 32;

o   On the uplink, the UE can be configured with HARQ mode A or B per HARQ process;

o   If a logical channel is configured with allowed HARQ-mode, it can only be mapped to a HARQ process with the same HARQ mode;

o   The value ranges of MAC (i.e. sr-ProhibitTimer and configuredGrant-Timer), RLC (i.e. t-Reassembly), and PDCP (i.e. discardTimer and t-reordering) layer timers are extended to accommodate the long propagation delay;

o   The Discontinuous Reception (DRX) timers drx-HARQ-RTT-TimerDL and drx-HARQ-RTT-TimerUL have been extended by the UE-gNB RTT.

– NTN-specific system information:

o   A new SIB, specifically SIB19, has been introduced for NTN which contains NTN-specific parameters related to the serving cell and/or neighbor cells, as defined in TS 38.331 [20].

– Mobility in idle mode (RRC_IDLE) and inactive mode (RRC_INACTIVE):

o   A Tracking Area corresponds to a fixed geographical area and any respective mapping is configured in the RAN;

o   The network can broadcast multiple TACs per PLMN in an NR-NTN cell, aimed at reducing the signaling load at cell edge; this applies in particular for Earth-moving cells. A TAC modification in the System Information is controlled by the network and may not be exactly synchronized with a real-time illumination of the on-ground beams;

o   Cell (re-)selection enhancements have been introduced in quasi-Earth fixed scenarios. In particular, the UE can implement time- and location-based measurements. The time/location information associated with a cell is provided via SI and they refer, respectively, to (i) the time when the serving cell is going to stop providing connectivity to the given geographical area; and (ii) the reference location and a distance threshold of a serving cell;

o   Location-assisted cell re-selection is supported in Quasi-Earth fixed deployments, based on the distance between the UE and the reference location of the cell (serving cell and/or neighboring cell);

o   Measurement rules for cell re-selection with timing and location information are specified in TS 38.304 [21].

– Mobility in connected mode (RRC_CONNECTED):

o   Radio Resource Management (RRM) Event D1 is a new measurement event based on location, which was introduced to trigger the location reporting based on UE's location;

o Also related to RRM measurements, the network can configure multiple Synchronization Signal (SS)/Physical broadcast channel (PBCH) Block Measurement Timing Configuration (SMTCs) in parallel per carrier and for a given set of cells depending on UE capabilities, using the propagation delay difference and ephemeris information. Moreover, it can configure measurement gaps based on multiple SMTCs;

o Adjusting SMTCs is possible under network control based on UE-assistance information if available for connected mode;

o UE supports mobility between NTN and TN (i.e. from NTN to TN and from TN to NTN), but is not required to connect to both NTN and TN at the same time. It may also support mobility between RAT based on different orbits (GSO, NGSO at different altitudes);

o To enable mobility in NTN, the network provides serving cell's and neighboring cell's satellite ephemeris needed to access the target serving NTN cell in the handover command;

o Triggering conditions upon which UE may execute Conditional Hand-Over (CHO) to a candidate cell, have been introduced: event A4 RRM measurement-based trigger, time-based trigger condition, and location-based trigger condition. The two last conditions are configured together with one of the measurement-based trigger conditions. Location is defined by the distance between UE and a reference location (Event D1). Time is defined by the time between T1 and T2, where T1 is an absolute time value and T2 is a duration started at T1.

- UE location reporting:

o For location reporting from the UE, following a network request and after the Access Stratum (AS) has been established in connected mode, a UE shall report a coarse location to the NG-RAN; the coarse location corresponds to the most significant bits of GNSS coordinates, which provide a 2 km accuracy in the location estimate.

With regard to NG-RAN architecture, and the specification of protocols for the related network interfaces, Release-17 introduced the following:

- The Cell Identity, indicated by the gNB to the Core Network as part of the User Location Information corresponds to a Mapped Cell ID, irrespective of the orbit of the NTN payload or the types of service links supported. It is used for Paging Optimization in NG interface, Area of Interest, and Public Warning Services.

- The Cell ID included within the target identification of the handover messages allows identifying the correct target radio cell as well as for RAN paging.

- The mapping between Mapped Cell IDs and geographical areas is configured in the RAN and Core Network. The gNB is responsible for constructing the Mapped Cell ID based on the UE location info received from the UE, if available. The mapping may be pre-configured (e.g., up to operator's policy) or up to implementation.
- The gNB reports the broadcasted TAC(s) of the selected PLMN to the Access and Mobility Management Function (AMF) as part of UE Location Information (ULI). In case the gNB knows the UE's location information, the gNB may determine the Tracking Area Indicator (TAI), the UE is currently located in and provide that TAI to the AMF as part of ULI.
- For a RRC_CONNECTED UE, when the gNB is configured to ensure that the UE connects to an AMF that serves the country in which the UE is located. If the gNB detects that the UE is in a different country than that served by the serving AMF, then it should perform an NG handover to change to an appropriate AMF, or initiate a UE Context Release Request procedure toward the serving AMF (in which case the AMF may decide to de-register the UE).
- The NTN-related parameters, as listed in clause 16.14.7 of TS 38.300 [22], shall be provided by O&M to the gNB providing non-terrestrial access. Additional NTN-related parameters in Annex B4 of TS 38.300 may be provided by O&M to the gNB for its operation.

With respect to Radio Frequency (RF) aspects, based on the adjacent channel coexistence studies reported in TR 38.863 [23], the minimum RF and performance requirements for NTN in FR1 have been defined and detailed in TS 38.101-5 [24]. These requirements apply to NR UE supporting satellite access as defined in TS 38.108 [25], for an NR SAN:

- The RF requirements of an NTN-enabled UE are the same as a UE operating inside a TN, which also allows connectivity to both network components;
- The RF requirements of the SAN, defined in TS 38.108, are lower compared to those of a TN base station, defined in TS 38.104 [26];
- Specific RRM requirements for the NTN component are reported in TS 38.133 [27], and they mainly relate to the maximum supported timing and frequency errors;
- Additionally to the SAN case, the RF requirements for HAPS were defined in TS 38.104 as a base station class referring to Wide Area base stations without additional modifications.
- NR operating band n1 can be applied for HAPS operation, as defined in TS 38.104;
- The NR UEs defined in TS 38.101-1 [28], can support HAPS without the need for additional modifications in the specification.

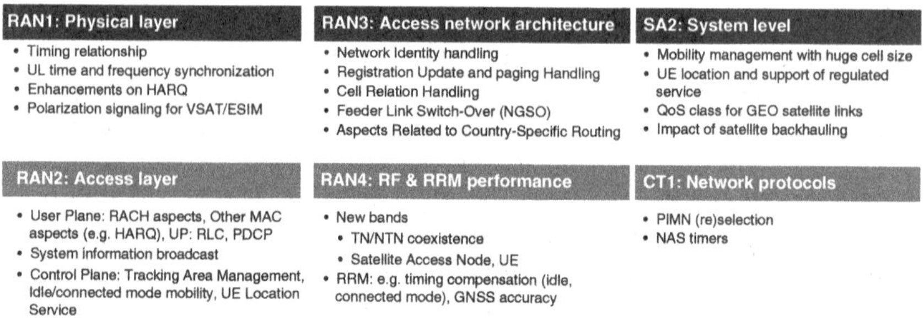

RAN1: Physical layer	RAN3: Access network architecture	SA2: System level
• Timing relationship • UL time and frequency synchronization • Enhancements on HARQ • Polarization signaling for VSAT/ESIM	• Network Identity handling • Registration Update and paging Handling • Cell Relation Handling • Feeder Link Switch-Over (NGSO) • Aspects Related to Country-Specific Routing	• Mobility management with huge cell size • UE location and support of regulated service • QoS class for GEO satellite links • Impact of satellite backhauling

RAN2: Access layer	RAN4: RF & RRM performance	CT1: Network protocols
• User Plane: RACH aspects, Other MAC aspects (e.g. HARQ), UP: RLC, PDCP • System information broadcast • Control Plane: Tracking Area Management, Idle/connected mode mobility, UE Location Service	• New bands • TN/NTN coexistence • Satellite Access Node, UE • RRM: e.g. timing compensation (idle, connected mode), GNSS accuracy	• PIMN (re)selection • NAS timers

Figure 1.12 3GPP Rel-17 NTN impacts on 3GPP specifications. Source: El Jaafari et al. [1].

In addition to the NR-NTN specifications, Release-17 also addressed NB-IoT and eMTC support over NTN, with a study performed in the framework of RAN WGs. Based on the outcomes of Release-17 analyses for NR-NTN and the Release-16 TR 38.821, the objective of this study was to identify the required features and adaptations to enable IoT-NTN in Release-17 with prioritized satellite access. Within the Work Item on "NB-IoT/eMTC support for NTN," Release-17 defines the enhanced features for E-UTRAN support of radio access via NTN for BL UEs, UEs in enhanced coverage, and NB-IoT UEs.

SA and CT aspects of NB-IoT/eMTC NTN in EPS provide minimum essential functionality for the Rel-17 UE and the network to support satellite E-UTRAN access in WB-S1 mode or NB-S1 mode with CIoT EPS optimization.

The support of IoT NTN is largely aligned with that of Rel-17 NR NTNs in 5GS, with the exception of discontinuous coverage that is addressed only within the IoT NTN work item in Rel-17.

Figure 1.12 summarized the impacts on 5G NR/NG-RAN specifications introduced by the support of NTN, at both RAN and SA levels.

At the time of writing this book, NTN Release-18 is addressing further enhancements for NR and NB-IoT/eMTC via NTN within the following Study and Work Items:

– Satellite-related SIs in SA:
  o 5G enhancements for satellite access-Phase 2, which includes enhancements in the architecture to support discontinuous coverage for mobility (for instance, paging enhancement) and other improvements considering prediction, awareness, and notification of the UE's wake-up time, and power saving optimization;
  o Enhanced Location Services (LCS), aiming at studying architectural enhancements for 5G with satellite access in collaboration with

RAN, in order to support network-verified UE location and network-controlled positioning. In addition, also the enhancements to meet the LCS requirements defined in TS 22.261 [9], are being addressed;
○ Satellite backhauling, addressing the architecture improvements for the support of backhaul with changing delay (e.g. brought by Inter-Satellite Links [ISLs] or changed satellite backhaul on the UP path), and/or limited bandwidth in conditions of a gNB with satellite backhaul only (e.g. restricted by the maximum data rate of a satellite beam);
– Satellite-related SIs and Work Items in RAN:
○ Study of the requirements and use cases for network-verified location of the UE via NTN [29]. The findings of this study are reported in 3GPP TR 38.822 [30];
○ Study and evaluate potential solutions for network to verify UE reported location information;
○ Definition of the enhancements aiming at [31]:
■ NR-NTN deployments above 10 GHz, to support new scenarios for fixed and mobile VSATs;
■ Coverage enhancement, focusing on the applicability of the general solutions developed for NR to NTN and on the identification of potential issues and enhancements taking into account the NTN peculiarities. This is expected to lead to optimized performance when considering handheld terminals, including smartphones;
■ Based on RAN WGs conclusions of the study phase, Release-18 is specifying necessary enhancements to allow the network to verify UE location which needs to be carried out in order to fulfil the regulatory requirements;
■ NTN-TN and NTN-NTN mobility and service continuity enhancements;
○ definition of the enhancements for IoT-NTN aiming at:
■ IoT-NTN Performance Enhancements in Rel-18 to address remaining issues from Rel-17. These include: Disabling of HARQ feedback to mitigate impact of HARQ stalling on UE data rates and specifying improved GNSS operations for a new position fix for UE pre-compensation during long connection times and for reduced power consumption.
■ improve mobility aspects such as support of neighbor cell measurements and corresponding measurements triggering before RLF, using Release-17 (TN) NB-IoT, eMTC as a baseline. As well as re-using the solutions introduced in Rel-17 NR NTN for mobility enhancements for eMTC, with minimum necessary changes to adapt them to eMTC.

- Study and specify, if needed, MM enhancements and power saving enhancements for discontinuous coverage, taking into account the conclusions from the SA2 study.
- Specify RF and RRM requirements for NB-IoT and eMTC operation over NTN functionality defined in the Rel-17.

Table 1.6 reports a list of the activities performed in Release-17 and -18 at RAN and SA level.

Table 1.7 summarize the 3GPP RAN and SA/CT specifications, respectively, that were updated to support NTN in the 5GS.

Table 1.6 List of satellite-related activities in 3GPP in the different TSG in Release-17 and Release-18.

3GPP Release	RAN	SA
17	• Specification of the solutions for 5G NR/NG-RAN to support NTN assuming a transparent payload architecture, FDD mode, and UEs with GNSS capabilities • Study the issues and specification of the solutions for NB-IoT/eMTC to support NTN access assuming a transparent payload architecture, FDD mode, and UEs with GNSS capabilities. The findings are captured in TR 36.763 [32]	• SA2: Specification of the solutions for the 5G architecture to support satellite access and satellite backhaul (fixed bandwidth and latency)
18	• Study of the requirements and use cases for network-verified UE location in NR-NTN. The findings are captured in TR 38.882 • Specification of the solutions for NR to support coverage enhancements in FR1, deployments above 10 GHz, network-verified UE location, and mobility enhancements • Specification of the solutions for NB-IoT/eMTC to optimize service performance, mobility, and to support discontinuous coverage	• SA1: Identification of the guidelines for extra-territorial 5G Systems such as satellite, reported in TR 22.926 [33]. • SA2: Define the solutions to support discontinuous coverage, backhaul with variable delay and/or limited bandwidth, and local switch for UEs in a communication when they are served by an on-board User Plane Function (UPF)

Table 1.7 3GPP RAN specifications updated for the integration of satellite components in the 5G.

3GPP Working group	3GPP RAN Specifications
RAN1	**TS 38.211**→Physical channels and modulation
	TS 38.213→Physical layer procedures for control
	TS 38.214→Physical layer procedures for data
RAN2	**TS 38.300**→Overall description; Stage-2
	TS 38.304→User Equipment (UE) procedures in idle mode and in RRC Inactive state
	TS 38.306→User Equipment (UE) radio access capabilities
	TS 38.321→Medium Access Control (MAC) protocol specification
	TS 38.322→Radio Link Control (RLC) protocol specification
	TS 38.323→Packet Data Convergence Protocol (PDCP) specification
	TS 38.331→Radio Resource Control (RRC); Protocol specification
RAN3	**TS 38.401**→NG-RAN; Architecture description
	TS 38.410→NG-RAN; NG general aspects and principles
	TS 38.413→NG-RAN; NG Application Protocol (NGAP)
	TS 38.423 NG-RAN; NG-RAN; Xn Application Protocol (XnAP)
RAN4	**TS 38.101-5**[b)]→User Equipment (UE) radio transmission and reception, part 5: Satellite access Radio Frequency (RF) and performance requirements
	TS 38.108→Satellite Access Node radio transmission and reception
	TS 38.133→Requirements for support of radio resource management
	TR 38.863→Non-terrestrial networks (NTN) related RF and coexistence aspects
	TS 38.104→Base Station (BS) radio transmission and reception
	TS 38.181→Satellite Access Node conformance testing; NTN specific characteristics

a) 3GPP RAN specifications updated for NB-IoT/eMTC support for NTN are within 36 series: Replace 38 per 36 in Table 1.7.
b) For IoT-NTN: Evolved Universal Terrestrial Radio Access (E-UTRA); User Equipment (UE) radio transmission and reception for satellite access are specified in TS 36.102.

1.11 NTN in Release-19 and Beyond

3GPP Release-19 content approved at the December 2023 TSGs (#102) defines the future direction of 5G-Advanced. The related 3GPP normative work will start in the first quarter of 2024. Based on the original RAN Release-19 timeline, as endorsed in RP-230050, December 2025 is the target date for ASN.1 freeze. The target stage 3 freeze in SA/CT is September 2025.

NTN continues to evolve in 3GPP Release-19 work adding new features and further enhancements at both SA and RAN levels. Overall, this Release aims at defining additional features for NTN, at improving the performance (e.g., throughput and radio link availability), and at providing new capabilities and features/network topology, e.g., leveraging regenerative payloads.

Within SA1, a new SI on satellite access (Phase 3) was agreed in May 2022, aiming at studying new use cases and the related regulatory requirements and to identify novel service requirements and enhancements. These include, for instance, the support of operations with intermittent or temporary NTN connectivity for delay-tolerant applications, GNSS-independent operation, enhancements to positioning, and communications between users covered by the same NTN node. The outcomes of the SA1 studies are being collected in TR 22.865 [34].

Table 1.8 reports the on-going activities in Release-19 for satellite access at SA1 and SA2 (see 3GPP SP-231199, SP-231802 and TR 23.700).

The features and further enhancements on NTN which are approved as part of the Release-19 content are listed in Tables 1.9 and 1.10 for NR NTN and IoT NTN, respectively.

Other candidate RF and RRM related features for NTN are expected to be further discussed in March and June 2024. Thereby, the Release-19 work plan for NTN may include the following additional features: High Power UE

Table 1.8 Release-19 enhancements on system architecture for the support of satellite access.

3GPP Release	SA
Rel.19	• SA1: Enhancements of the 5GS over satellite, including operation with intermittent/temporary satellite connectivity for delay-tolerant communication service, GNSS-independent operation, and communication between UEs under the same satellite's coverage • SA2: Support of regenerative payload with Store and forward capabilities for IoT-NTN and UE-Satellite-UE mesh connectivity capability for NR-NTN

Table 1.9 Further enhancements/features for NR NTN in Release-19.

3GPP Release-19 NR NTN content approved at the December 2023 TSGs (RAN#102)

- Coverage enhancements for downlink, at both link level and system level
- Uplink capacity/throughput enhancement with enhanced multiplexing techniques on the uplink
- Support for regenerative payloads (at least gNB on board)
- Specify signalling of the intended service area of a broadcast service (e.g., MBS broadcast) via NR NTN
- Support of UEs with reduced capabilities (Rel-17 RedCap and Rel-18 eRedCap UEs) with NR NTN operating in FR1-NTN bands

---

Table 1.10 Further enhancements/features for IoT NTN in Release-19

3GPP Release-19 IoT NTN content approved at the December 2023 TSGs (RAN#102)

- Support of regenerative payloads with Store and Forward Satellite operation (e.g., eNB with switching and edge computing capability on-board) to support delay-tolerant services in a lacunar (discontinuous coverage) NGSO constellation with a reduced number of gateways
- Uplink capacity/throughput enhancement with enhanced multiplexing techniques on the uplink

---

(HPUE) in FR1 bands, mobile VSAT UE for NGSO in above 10 GHz bands, enhanced channel model for dynamic behaviour testing, and the support of new bands for NR-NTN, such as Ku-band and C-band.

1.12 3GPP and Standardization

The 3rd Generation Partnership Project (3GPP) gathers together seven Standard Development Organizations (SDO) and is the global leader in the definition of cellular communication standards. Such Organizational Partners, coming from Asia, Europe, and North America, are ARIB, ATIS, CCSA, ETSI, TSDSI, TTA, and TTC (Figure 1.13). 3GPP provides to all its members a stable and common environment for the definition of TR and Specifications that cover cellular telecommunication technologies, including radio access, core network, and service capabilities, providing a unified and complete description of the technologies for mobile telecommunications.

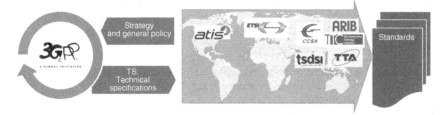

Figure 1.13 3GPP and standardization.

Since the first 3GPP meeting in December 1988, as of December 2022, 3GPP produced 17 Releases defining the global standards for 2G, 3G, 4G, and now, 5G mobile networks. The three TSG are RAN, Service and SA, and Core network and Terminals (CT). The work method adopted by 3GPP TSGs, including their WG and sub-groups, are defined in TR 21.900 [35].

Table 1.11 provides a list of RAN, SA, and CT WGs, and those relevant to the activities on NTN are highlighted in gray.

To better understand the massive impact that NTN had since its initial introduction in 3GPP, and to clearly highlight the huge joint effort of satellite and terrestrial stakeholders, Figure 1.14 shows an estimate of the number of

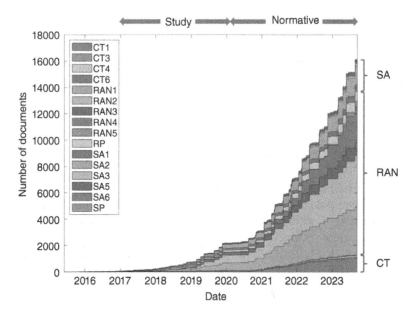

Figure 1.14 Estimated number of Technical Documents submitted to the 3GPP TSGs, per WGs.

Table 1.11 Relevant 3GPP working groups for Non-Terrestrial Networks.

RAN WGs	objective
RAN1: Radio Layer 1 specification	compatibility of 5G physical layer for operation via satellite
RAN2: Radio Layer 2 and Radio Layer 3 RR specification	compatibility of 5G access layer for operation via satellite
RAN3: Iub Iur and Iu specification - UTRAN O&M requirements	compatibility of NG protocols/architecture with satellite access/transport
RAN4: Radio performance and protocol aspects (system) - RF parameters and BS conformance	performance of NG-RAN via satellite
RAN5: Mobile terminal conformance testing	conformance test definitions
RAN6: Legacy RAN radio and protocol	Not relevant
RAN AHG1: Ad-hoc group on ITU (internal) co-ordination	Liaisons between 3GPP and ITU

SA WGs	objective
SA1: Services	5G service requirements covering satellite use cases
SA2: Architecture	architecture requirements covering satellite network
SA3: Security	compatibility of the security architecture with integrated satellite scenarios
SA4: Codec	Not relevant
SA5: Telecom Management	compatibility of the network management architecture with integrated satellite scenarios
SA6: Mission-critical applications	extension for satellite

CT WGs	objective
CT1: MM/CC/SM [Iu]	network protocol able to cope with extended latency
CT3: Interworking with external networks	Not relevant
CT4: MAP/CAMEL/GTP/BC H/SS	Not relevant
CT6: Smart Card Application Aspects	Not relevant

Technical Documents submitted to 3GPP meetings between January 2016 and September 2023. Between 2016 and 2017, less than 100 contributions were submitted related to satellites, while, as of September 2023, more than 16 000 documents were presented.[6]

References

1 M. El Jaafari, N. Chuberre, S. Anjuere, L. Combelles, "Introduction to the 3GPP-defined NTN standard: a comprehensive view on the 3GPP work on NTN", *International Journal of Satellite Communications and Networking*, vol. 41, no. 3, pp. 220–238, 2023. doi: 10.1002/sat.1471.

2 M. El Jaafari, N. Chuberre, "Guest editorial IJSCN special issue on 3GPP NTN standards for future satellite communications", *International Journal of Satellite Communications and Networking*, vol. 41, no. 3, pp. 217–219, 2023. doi: 10.1002/sat.1472.

3 3GPP TR 38.811: "Study on New Radio (NR) to support non-terrestrial networks", October 2020.

4 3GPP TR 22.822: "Study on using satellite access in 5G", July 2018.

5 ITU-R Resolution 65, "Principles for the process of future development of IMT for 2020 and beyond", 2015.

6 ITU-R Report M.2514-0, "Vision, requirements and evaluation guidelines for satellite radio interface(s) of IMT-2020", September 2022.

7 ITU-R Recommendation M.2083, "IMT Vision – Framework and overall objectives of the future development of IMT for 2020 and beyond", September 2015.

8 A. Guidotti et al., "Role and Evolution of Non-Terrestrial Networks towards 6G systems", to be submitted to *IEEE Access*, October 2023.

9 3GPP TS 22.261: "Service Requirements for the 5G System", June 2023.

10 3GPP TR 38.821: "Solutions for NR to support Non-Terrestrial Networks (NTN)", April 2023.

11 ITU-R, World Radiocommunication Conference 2023 (WRC-23) Provisional Final Acts, December 2023. Available: https://www.itu.int/wrc-23/

12 CEPT Weekly Reports from WRC-23, November–December 2023. Available: https://cept.org/ecc/groups/ecc/cpg/page/weekly-report-from-wrc-23

13 3GPP, SP-180997: "Stage 1 CRs on SMARTER", December 2018.

---

6 This information was obtained by means of a web-scraping tool developed at the University of Bologna and not from official 3GPP sources. As such, the exact numbers might be subject to variations, but the general trends are nonetheless representative.

14 3GPP TR 23.737: "Study on architecture aspects for using satellite access in 5G", March 2021.

15 3GPP TR 28.808: "Study on management and orchestration aspects of integrated satellite components in a 5G network", April 2021.

16 3GPP TS 23.501: "System architecture for the 5G System (5GS)", September 2023.

17 3GPP TS 24.821: "Study on PLMN selection for satellite access", September 2021.

18 3GPP RP-193234: "Solutions for NR to support non-terrestrial networks (NTN)", December 2019.

19 3GPP TS 38.321: "Medium Access Control (MAC) protocol specification", June 2023.

20 3GPP TS 38.331: "NR; Radio Resource Control (RRC); Protocol specification", July 2023.

21 3GPP TS 38.304: "NR; User Equipment (UE) procedures in idle mode and in RRC Inactive state", July 2023.

22 3GPP TS 38.300: "NR; NR and NG-RAN Overall description; Stage-2", June 2023.

23 3GPP TR 38.863: "Non-terrestrial networks (NTN) related RF and co-existence aspects", April 2023.

24 3GPP TS 38.101-5: "NR; User Equipment (UE) radio transmission and reception; Part 5: Satellite access Radio Frequency (RF) and performance requirements", June 2023.

25 3GPP TS 38.108: "NR; Satellite Access Node radio transmission and reception", June 2023.

26 3GPP TS 38.104: "NR; Base Station (BS) radio transmission and reception", June 2023.

27 3GPP TS 38.133: "NR; Requirements for support of radio resource management", June 2023.

28 3GPP TS 38.101-1: "NR; User Equipment (UE) radio transmission and reception; Part 1: Range 1 Standalone", June 2023.

29 3GPP RP-221820: "New SID: Study on requirements and use cases for network verified UE location for Non-Terrestrial-Networks (NTN) in NR", June 2022.

30 3GPP TR 38.882: "Study on requirements and use cases for network verified UE location for Non-Terrestrial-Networks (NTN) in NR", June 2022.

31 3GPP RP-223534: "Revised WID: NR NTN (Non-Terrestrial Networks) enhancements", December 2022.

32 3GPP TR 36.763: "Study on Narrow-Band Internet of Things (NB-IoT) / enhanced Machine Type Communication (eMTC) support for Non-Terrestrial Networks (NTN)", June 2021.

33 3GPP TR 22.926: "Guidelines for extraterritorial 5G Systems (5GS)", December 2021.

34 3GPP TR 22.865: "Study on satellite access - Phase 3", June 2023.

35 3GPP TR 21.900: "Technical Specification Group working methods", September 2023.

2

The 3GPP 5G Overview

2.1 Introduction

Previous generations of mobile telecommunications technologies (1G to 4G) addressed the demand for mobile voice and broadband data; 5G enables new services, ecosystems, and revenues on top of the existing ones, while also delivering improved end-user experience for existing scenarios. Possible use cases for 5G have been classified by the International Telecommunications Union (ITU) as Enhanced Mobile Broadband (eMBB), Ultra-Reliable Low Latency Communications (URLLC), and Massive Machine-Type Communications (mMTC) [1]. eMBB is the natural evolution of mobile broadband services, including augmented and virtual reality, 3D videos, 4K streaming, and others – these are still the largest share of services delivered through mobile telecommunications today. URLLC services require ultra-high reliability and low latency: they include wireless connection and control of industrial manufacturing, machinery, and production processes, intelligent transport systems, traffic safety, and control. mMTC services are characterized by a massive number of devices (the so-called IoT), addressing applications such as smart cities, smart homes, utilities, remote monitoring, and others. However, 5G may also address many other scenarios and applications, such as Fixed Wireless Access (FWA) (where 5G is used as last mile access, complementing other technologies) and others that can only be envisaged today.

The above use cases resulted in a mix of very different (and possibly even contrasting) requirements, but the 5G system architecture has been specified from the ground up to support such a diverse mix of requirements in a flexible manner. This is thanks to the standardization process in the 3rd Generation Partnership Project (3GPP), which makes it possible to specify the best technical solution under the given constraints and requirements

5G Non-Terrestrial Networks: Technologies, Standards, and System Design, First Edition.
Alessandro Vanelli-Coralli, Nicolas Chuberre, Gino Masini, Alessandro Guidotti, and Mohamed El Jaafari.
© 2024 The Institute of Electrical and Electronics Engineers, Inc. Published 2024 by John Wiley & Sons, Inc.

from the interested parties (operators, equipment vendors, chipset makers, and others).

2.2 5G System Architecture

An in-depth description of the 5G system architecture for New Radio (NR) is beyond the scope of this book. We will only focus on the main concepts, which are necessary to understand the role of Non-Terrestrial Networks (NTN) in 5G.

As with earlier generations of mobile telecommunications systems, the components of a 5G system are the core network (5GC), the New (G) RAN (NG-RAN), and the user equipment (UE).

The 5G system architecture as specified by 3GPP is composed of *logical nodes*. A logical node in general can be characterized as follows [2]:

- It is defined as a collection of *logical functions*, described in a "Stage 2" type of Technical Standard (TS). For 5GC, such a description is in [3, 4], and for NG-RAN, it is in [5–7].
- It terminates a set of *logical interfaces* with other logical nodes.
- The logical interfaces it terminates are defined based on the logical functions in the node.
- At least in the Radio Access Network (RAN), logical interfaces are defined starting from their physical layer, transport requirements, and protocols.
- According to different implementations, logical nodes may be deployed or grouped in different ways. The logical architecture does not mandate any specific deployment (e.g., co-located, virtualized, and centralized), but rather it accommodates such deployments. Two logical nodes, for example, might be implemented as a single physical entity and still comply with the standardized logical architecture: in this case, any logical interface between them would become internal to the physical entity and thus externally "invisible" in the implementation.

2.2.1 5G Core Network

The functions in the 5GC include [8]:

- *Storing subscription information*: identifiers, security information needed to authenticate UEs and derive ciphering keys, allowed/restricted networks, Radio Access Technology (RATs), geographic areas for a given UE, and others.
- *Authenticating and authorizing the UE* toward the network and vice versa.

- *Registering the UE* and keeping track of the list of registered UEs.
- *Tracking UE location* with different granularity levels depending on the UE state (e.g. active, and idle).
- *Establishing data sessions* on different data networks following the UE request, according to the payload type (e.g. IPv4, IPv6, and Ethernet).
- *Forwarding traffic* between the RAN and the data networks that a UE has established sessions to.
- *Deriving and enforcing Quality of Service (QoS) and charging rules* according to the operator's policies (the enforcement is performed together with the RAN).
- *Lawful interception* (providing metadata and access to target payload) according to legal obligations.

2.2.2 NG Radio Access Network

The building block for NG-RAN is the NG-RAN node. It can be either a gNB (a base station providing NR radio access) or an ng-eNB (a base station providing 4G radio access, with some added enhancements). NG-RAN nodes are connected to the 5GC with the NG interface and to one another with the Xn interface [9–21].

The functions in the gNB and ng-eNB include [6]:

- *Radio resource management*, including radio bearer control, radio admission control, connection mobility control, and dynamic allocation of resources to UEs in both Uplink (UL) and Downlink (DL).
- *Header compression, data encryption, and integrity protection.*
- *User Plane (UP) and Control Plane (CP) routing.*
- *Connection setup and release.*
- *Scheduling and transmission of system broadcast information* and *paging messages.*
- *Measurement and reporting configuration for mobility and scheduling.*
- *Support for network slicing.*
- *QoS flow management and mapping* to data radio bearers.

The NG-RAN logical architecture is shown in Figure 2.1. It accommodates both types of NG-RAN nodes and supports the capability of NR to operate in both *Stand-Alone (SA)* and *Non-Stand-Alone (NSA)* modes.

In SA operation, gNBs connect to Access and Mobility Functions (AMFs) and User Plane Functions (UPFs) (both RAN and core network are "native" 5G). In NSA operation, gNBs and ng-eNBs tightly interoperate with one another, connecting to an existing 4G core network, to provide *dual connectivity (DC)* toward UEs for enhanced data rates [7].

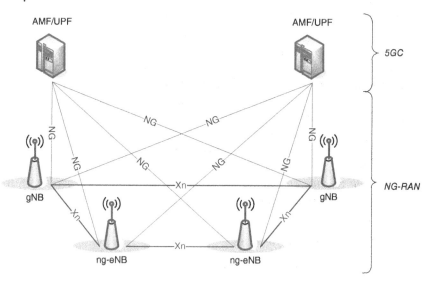

Figure 2.1 The NG-RAN logical architecture [6].

2.2.3 Dual Connectivity

Dual connectivity in NG-RAN is conceptually similar to the functionality specified by 3GPP for Enhanced Universal Terrestrial RAN (E-UTRAN) as far back as Rel-13: a UE capable of multiple reception and transmission is configured to use resources provided by two different nodes. One node takes on the role of *Master Node* (MN) and the other of *Secondary Node* (SN). The MN is connected to the core network and "owns" the UE connection; the MN and the SN are connected via a direct network interface (Xn in NG-RAN, or X2 in E-UTRAN) [2]. It is generally assumed that the MN is a macro base station providing coverage while the SN is a smaller cell providing additional capacity, although other types of deployments are also supported [8].

From a CP perspective, the UE in DC has a single state in the Radio Resource Control (RRC) protocol, based on the MN RRC and a single control plane connection toward the core network. Each RAN node has its own RRC entity (Enhanced Universal Terrestrial Radio Access [E-UTRA] RRC if it is an eNB and NR RRC if it is a gNB), which can generate RRC packets (RRC Protocol Data Units [PDUs]) to be sent to the UE. RRC PDUs generated by the SN can be transported via the MN to the UE [2].

The MN always sends the initial SN RRC configuration to the UE, but subsequent reconfigurations may be transported via MN or SN.

When transporting RRC PDUs from the SN, the MN does not modify the UE configuration provided by the SN. Split Dedicated Radio Bearers (split DRBs) are supported, allowing duplication of RRC PDUs generated by the MN via the direct path and via the SN [2].

From a UP perspective, three bearer types exist: Master Cell Group (MCG) bearer, Secondary Cell Group (SCG) bearer, and split bearer [2].

2.2.4 Connectivity Options

Because of its dual-faceted nature, which accommodates both existing and upcoming network deployments, an analogy with the ancient Roman god *Janus Bifrons* has been used to describe the NG-RAN architecture [2]. Depending on the role that gNBs and eNBs may take for dual connectivity (as *master RAT* or *secondary RAT*), and depending on the core network type, a number of different alternatives (*connectivity options*, see Table 2.1) have been specified [7], with the ambition level of supporting various "migration paths" for operators toward 5G.

Not all architecture options, though, seem justified by real-life requirements. For example, Option 3 (E-UTRAN-NR Dual Connectivity [EN-DC]) is arguably the best short-term option for 5G deployment as it relies on an existing Evolved Packet Core (EPC) and uses the E-UTRAN network interfaces S1 and X2. Option 2 (SA), on the other hand, requires the 5GC to enable the delivery of advanced and distinctive 5G services, including URLLC, hence it seems to be the preferable long-term target option. Indeed, for the most part, the industry seems to have decided to base the deployment of 5G on Options 3 and 2 [22].

Table 2.1 Connectivity options [22]. Option 2 also supports NR as both Master and Secondary RAT for DC (NR-NR DC).

Connectivity option	Core network	Master RAT	Secondary RAT	3GPP term	3GPP release
Option 1	EPC	LTE	—	LTE	Rel-8
Option 3	EPC	LTE	NR	EN-DC	Rel-15, 12/2017
Option 2	5GC	NR	—	NR	Rel-15, 6/2018
Option 4	5GC	NR	eLTE	NE-DC	Rel-15, 3/2019
Option 5	5GC	eLTE	—	eLTE	Rel-15, 6/2018
Option 7	5GC	eLTE	NR	NGEN-DC	Rel-15, 3/2019

Source: Cagenius et al. [22]/Telefonaktiebolaget LM Ericsson.

2.2.5 Split Architecture

In order to support the whole set of diverse 5G use cases, NG-RAN supports placing selected logical functions closer to the network edge and to separate CP and UP handling as needed by the operator. Distributing RAN functions between a "centralized" and a "distributed" unit is also supported [2]. Some of the benefits of such capabilities include [21, 23]:

- More flexibility in hardware implementation, allowing better scalability and cost-effectiveness.
- Better coordination of features and load management, and better performance optimization.
- Enabling virtualized deployments and software-defined networking.
- Better adaptation to different user density and load demand in a given geographical area.
- Better adaptation to variable transport network performance.

Because of the above, NR supports the following:

- Splitting the previously *monolithic* RAN node.
- Such a split is standardized.
- *Monolithic* and *split* gNBs coexist and interoperate, and the rest of the network ignores whether a certain gNB is split or not.

The split architecture, then, remains nested inside the gNB, but it is fully specified in the standard to ensure interoperability. The *Matryoshka* (traditional Russian nested wooden dolls) has been used as an analogy: the outer *Matryoshka* always appears the same on the outside regardless of its contents, but it can be opened to expose the inner *Matryoshka(s)*, if present [2].

2.2.5.1 CU–DU Split

The gNB may be split into a gNB-*central unit* (CU) and a gNB-*distributed unit* (DU). The gNB-CU and the gNB-DU are both logical nodes, connected by the F1 interface (which has both CP and UP functions, defined in [20, 24–28]). Figure 2.2 shows the NG-RAN architecture with the split gNB [5]. The gNB-CU terminates the NG and Xn interfaces toward the rest of the network and hosts the RRC, Service Data Adaptation Protocol (SDAP), and Packet Data Convergence Protocol (PDCP) (high layers) and may connect to one or more gNB-DUs. The gNB-DU hosts the Radio Link Control (RLC), Medium Access Control (MAC), and PHYsical (PHY) layers (low layers) and may support several cells (but one cell is supported by only one gNB-DU). A gNB-DU connects to only one gNB-CU [5].

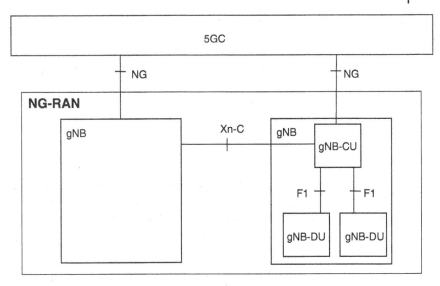

Figure 2.2 NG-RAN architecture with a split gNB [5].

The gNB-CU manages the UE context (the set of information associated with a UE served by the gNB) and requests the gNB-DU to allocate or modify the required radio resources for the UE. The gNB-DU accepts or rejects the request depending on whether the requested radio resources are available (and it can also request to modify the resources for an existing UE context). The gNB-DU is also responsible for scheduling and broadcasting system information and for encoding the New Radio Master Information Block (NR MIB) and System Information Block 1 (SIB1); the gNB-CU encodes the other System Information messages [24, 27]. The functional split between gNB-CU and gNB-DU ("centralized" PDCP and RRC; "decentralized" RLC, MAC, and PHY) closely resembles the protocol stack for DC [2].

Because of the above architecture choices, the gNB-CU and gNB-DU are always part of the same gNB once F1 is up and running. Disconnecting a gNB-DU from its gNB-CU and reconnecting it to a different one will "break" the gNB and create a new one: all existing information (including node configuration and all UE contexts currently handled) will be lost, and all connections will be dropped. In other words, unlike the NG interface for which "NG-flex" configuration is supported (each NG-RAN node is connected to all AMFs of AMF Sets within an AMF Region supporting at least one of the slices also supported by the NG-RAN node [5]), no such functionality is possible for F1. The only possibility for a gNB-DU to be connected to multiple gNB-CUs (e.g., for resiliency) is through appropriate proprietary implementation [5] (hence, the standard does not describe such an option).

The above functional distribution allows the centralization of traffic aggregation from several transmission points and the further separation of the PDCP for UP and CP into different central entities. This allows a separate UP handler while having a centralized RRC, increasing scalability according to UP traffic load. All of this ultimately resulted in the specification of the CP–UP split, further opening up the gNB-CU *Matryoshka* [2].

2.2.5.2 CP–UP Split

By splitting the PDCP into its UP and CP parts into different central entities, it is possible to further optimize the location of the different RAN functions. Other advantages of separating CP and UP handling (a concept that has been established for quite some time in telecommunication networks) include [8]:

- The flexibility to manage complex networks supporting different transport network topologies.
- The ability to tailor the deployment for various service requirements.
- Independent scaling of CP and UP, adding dedicated resources when needed.
- Support for multi-vendor interoperability between CP and UP nodes.

Splitting the gNB-CU results in the gNB-CU-CP and gNB-CU-UP logical nodes, connected by the E1 interface (which only has CP functions [29–32]).

The gNB-CU-CP hosts RRC and the CP part of the PDCP protocol; it also terminates the CP part of F1 (F1-C) toward the gNB-DU, as well as E1 toward the gNB-CU-UP. The gNB-CU-UP hosts SDAP and the UP part of PDCP; it also terminates the UP part of F1 (F1-U) toward the gNB-DU, as well as E1 toward the gNB-CU-CP [5].

If split into its CP and UP parts, a gNB may consist of one gNB-CU-CP, one or more gNB-CU-UPs, and one or more gNB-DUs. A gNB-CU-UP is connected to only one gNB-CU-CP, but implementations allowing a gNB-CU-UP to connect to multiple gNB-CU-CPs (e.g., for redundancy) are not precluded. One gNB-DU can connect to multiple gNB-CU-UPs under the control of the same gNB-CU-CP, and one gNB-CU-UP can connect to multiple gNB-DUs under the control of the same gNB-CU-CP [5] (Figure 2.3).

The gNB-CU-CP requests the gNB-CU-UP to set up, modify, and release the *bearer context*; the gNB-CU-UP accepts or rejects the request depending on whether the requested resources are available. The gNB-CU-UP can also request a bearer context modification to the gNB-CU-CP [5, 29, 32].

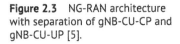

Figure 2.3 NG-RAN architecture with separation of gNB-CU-CP and gNB-CU-UP [5].

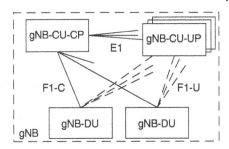

Among other things, the gNB-CU-CP is responsible for bearer mapping configuration and security configuration (which it then provides to the gNB-CU-UP), and for notifying the gNB-CU-UP of suspension and resumption of bearer contexts (typically in conjunction with a UE entering and leaving idle state). The gNB-CU-UP is responsible for notifying the gNB-CU-CP of user data inactivity, so that gNB-CU-CP may take further action, and for reporting data volume to the gNB-CU-CP [5].

2.2.6 IoT and MTC Aspects

The IoT is part of a transformation affecting industries, consumers, and the public sector. It provides the capability to embed electronic devices into "smart" objects that allow interaction with the physical world through sensing and actuation, enabling networking among smart objects, applications, and servers. The 5G system architecture addresses the requirement categories for Machine-Type Communications (MTC), essentially addressing the IoT: *Massive MTC* (mMTC) and *Critical MTC* (cMTC) [33].

Massive MTC addresses large numbers of simple devices, which require small and infrequent data transfers, to be massively deployed. Ubiquity of deployment, scalability to many connected devices, and low cost then motivate ultra-low complexity devices powered by non-rechargeable batteries for years. Examples of mMTC use cases are utility metering and monitoring, vehicle fleet management, industrial sensors, inventory and asset tracking for logistics, and others. Critical MTC addresses demanding IoT use cases requiring very high reliability and availability, and very low latency. Examples of cMTC use cases are remote or autonomous driving, cooperative safety, distribution automation in a smart grid, remote or automated vehicles in manufacturing or mining operations, and others. The cMTC requirement category is also referred to as URLLC [33].

Starting from 3GPP Rel-16, the 5G system supports IoT and MTC through the standardized architecture. A number of optimizations specific to cellular IoT, including eDRX and PSM, have been added to prepare the 5GC for IoT [33]. NG-RAN supports IoT: an NB-IoT UE is only supported by an ng-eNB [6].

2.3 3GPP and 5G Standardization

Standardization makes it possible to specify the best possible technical solution given all the requirements of interested parties (especially operators, network infrastructure vendors, and chipset makers).

The 3GPP provides the Reports and Specifications that define 3GPP technologies. The project covers cellular telecommunications technologies, including radio access, core network, and service capabilities, which provide a complete system description for mobile telecommunications. 3GPP specifications also provide hooks for non-radio access to the core network, and for interworking with non-3GPP networks. 3GPP specifications and studies are contribution-driven, by member companies, in Working Groups (WGs) and at the Technical Specification Group (TSG) level. The three TSGs in 3GPP are as follows [34] (Figure 2.4):

- Radio Access Networks (RAN),
- Services and Systems Aspects (SA),
- Core Network and Terminals (CT).

The WGs, within the TSGs, meet regularly and come together for the quarterly TSG plenary meeting, where their work is presented for information, discussion, and approval. TSG SA is also responsible for the overall coordination of the technical work and for monitoring its progress [34].

From LTE onwards, 3GPP has become the focal point for the vast majority of mobile systems beyond 3G. The progress of 3GPP standards is indicated by the milestones achieved in particular *Releases*. A release identifies a collection of functionalities, providing developers with a stable platform for the implementation of features at a given point and allowing the addition of new functionality in subsequent Releases [35]. New features are *functionality-frozen* and are ready for implementation when a Release is completed [34].

The major focus for all 3GPP Releases is to make the system backward and forward compatible as much as possible, ensuring uninterrupted operation of UE [36]. For example, while specifying NSA ("early drop" of 5G NR), "forward compatibility" was built into NSA NR equipment, to ensure that it would work with SA 5G NR systems [34].

Current NG-RAN architecture has been specified by 3GPP RAN WG3 over a period of 2 years (2017–2019, see Table 2.1) starting from a mix of very

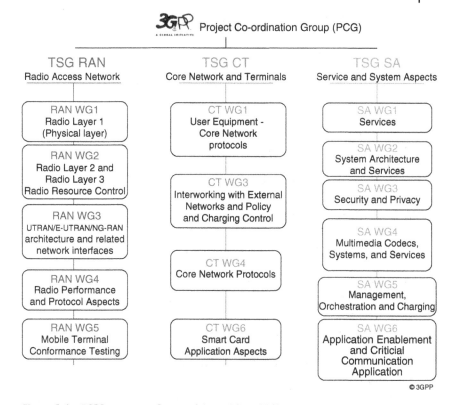

Figure 2.4 3GPP structure. Source: Adapted from [34].

different (and often contrasting) requirements, under intense time pressure, while the industry was still consolidating and managing the success of LTE. The level of flexibility achieved could have resulted in a considerable cost in terms of complexity. But thanks to the hard work and ingenuity of the 3GPP community the enormous effort to standardize and maintain the NG-RAN architecture has been successful [2].

References

1 ITU-R M.2083, IMT vision-framework and overall objectives of the future development of IMT for 2020 and beyond, ITU-R Rec. M02083, Setember 2015.
2 X. Lin, N. Lee, eds., *5G and Beyond – Fundamentals and Standards*, Springer, 2021.

3 3GPP TS 23.501: "System architecture for the 5G System (5GS); Stage 2", v. 17.1.1.
4 3GPP TS 23.502: "Procedures for the 5G System (5GS); Stage 2", v. 17.1.0.
5 3GPP TS 38.401: "NG-RAN; Architecture description", v. 15.6.0.
6 3GPP TS 38.300: "NR; Overall description; Stage-2", v. 15.7.0.
7 3GPP TS 37.340: "NR; Multi-connectivity; Overall description; Stage-2", v. 15.7.0.
8 S. Sirotkin, ed., *5G Radio Access Network Architecture – The Dark Side of 5G*, Wiley, 2021.
9 3GPP TS 38.410: "NG-RAN; NG general aspects and principles", v. 15.2.0.
10 3GPP TS 38.411: "NG-RAN; NG layer 1", v. 15.0.0.
11 3GPP TS 38.412: "NG-RAN; NG signalling transport", v. 15.3.0.
12 3GPP TS 38.413: "NG-RAN; NG Application Protocol (NGAP)", v. 15.5.0.
13 3GPP TS 38.414: "NG-RAN; NG data transport", v. 15.2.0.
14 3GPP TS 38.415: "NG-RAN; PDU Session User Plane protocol", v. 15.2.0.
15 3GPP TS 38.420: "NG-RAN; Xn general aspects and principles", v. 15.2.0.
16 3GPP TS 38.421: "NG-RAN; Xn layer 1", v. 15.1.0.
17 3GPP TS 38.422: "NG-RAN; Xn signalling transport", v. 15.3.0.
18 3GPP TS 38.423: "NG-RAN; Xn Application Protocol (XnAP)", v. 15.5.0.
19 3GPP TS 38.424: "NG-RAN; Xn data transport", v. 15.2.0.
20 3GPP TS 38.425: "NG-RAN; NR user plane protocol", v. 15.6.0.
21 B. Bertenyi, R. Burbidge, G. Masini, S. Sirotkin, Y. Gao, "NG radio access network (NG-RAN)", *Journal of ICT Standardization*, vol. 6, no. 1–2, pp. 59–76, January–May 2018.
22 T. Cagenius, A. Ryde, J. Vikberg, P. Willars, "Simplifying the 5G ecosystem by reducing architecture options", *Ericsson Technology Review*, November 30, 2018. Available: https://www.ericsson.com/en/ericsson-technology-review/archive/2018/simplifying-the-5g-ecosystem-by-reducing-architecture-options.
23 3GPP TR 38.801: "Study on new radio access technology: radio access architecture and interfaces," v. 14.0.0.
24 3GPP TS 38.470: "NG-RAN; F1 general aspects and principles", v. 15.6.0.
25 3GPP TS 38.471: "NG-RAN; F1 layer 1", v. 15.0.0.
26 3GPP TS 38.472: "NG-RAN; F1 signalling transport", v. 15.5.0.
27 3GPP TS 38.473: "NG-RAN; F1 Application Protocol", v. 15.7.0.
28 3GPP TS 38.474: "NG-RAN; F1 data transport", v. 15.3.0.
29 3GPP TS 38.460: "NG-RAN; E1 general aspects and principles", v. 15.4.0.
30 3GPP TS 38.461: "NG-RAN; E1 layer 1", v. 15.1.0.
31 3GPP TS 38.462: "NG-RAN; E1 signalling transport", v. 15.5.0.

32 3GPP TS 38.463: "NG-RAN; E1 Application Protocol (E1AP)", v. 15.5.0.

33 O. Liberg, M. Sundberg, E. Wang, J. Bergman, J. Sachs, G. Wikström, *Cellular Internet of Things – From Massive Deployments to Critical 5G Applications*, 2nd ed., Academic Press, 2019.

34 "About 3GPP", https://www.3gpp.org/about-3gpp.

35 "Releases", https://www.3gpp.org/specifications/67-releases.

36 B. Bertenyi, W. Chen, R. Burbidge, G. Masini, X. Zhou, J. John, *Interoperability and Compatibility of 5G Specifications*, November 27, 2018. Available: https://www.3gpp.org/news-events/1994-copatibility.

3

Non-Terrestrial Networks Overview

3.1 Elements of a Satellite Communications System

Satellite Communications encompass the use of an artificial satellite to establish communication with different locations on or close to the Earth's surface. Thus, a Satellite Communications system, or Non-Terrestrial Network (NTN), can be defined as a telecommunications system, including a communication satellite in its end-to-end link. As shown in Figure 3.1, a Satellite Communications system is composed of several segments:

- The *Space/Aerial segment*, which includes one or more satellites High-Altitude Platform Stations (HAPS). In general, we can refer to the space segment when the former case is considered and to the aerial segment in the latter. The satellites in the space segment are typically organized in a constellation orbiting around the Earth. The network elements in the space segment perform all the communication functions in the sky. Each element in the space/aerial segment includes: (i) a payload, i.e., the receiving/transmitting antennas and all the electronic equipment supporting the telecommunications operations; and (ii) a *platform*, i.e., all of the subsystems allowing the payload to perform its operations. The payload can further be classified as: (i) *transparent*, which basically acts as a radio repeater by implementing filtering, frequency conversion, and amplification operations; and (ii) *regenerative*, when the satellite/HAPS is equipped with an On-Board Processor (OBP) that allows to perform advanced operations, e.g., modulation/demodulation, Forward Error Correction (FEC).
- The *Control Segment*, which is composed of all of the on-ground elements aimed at monitoring and tracking the elements in the space/aerial segments and at guaranteeing the proper functioning of the overall

5G Non-Terrestrial Networks: Technologies, Standards, and System Design, First Edition.
Alessandro Vanelli-Coralli, Nicolas Chuberre, Gino Masini, Alessandro Guidotti, and Mohamed El Jaafari.

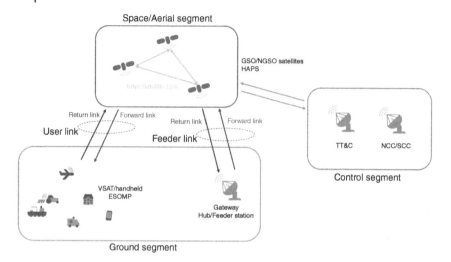

Figure 3.1 Segments and links of a general Non-Terrestrial Network (NTN) system.

system. For instance, this segment includes: (i) Telemetry Tracking and Control (TT&C), which tracks the position of the satellite(s)/HAPS(s) and telemeters and commands the on-board functions and (ii) the NTN Control Center (NCC) or System Control Center (SCC), which performs several operations, including the user configuration management, capacity management, channel allocation, and user access management.

- The *Ground Segment*, consisting of all of the Earth stations communicating with each other through the services provided by the space/aerial and control segments. These stations can be quite heterogeneous, but in general include: (i) the user terminals, which, depending on their size and characteristics, can be Very Small Aperture Terminals (VSAT), handheld terminals, or Earth Station on Moving Platform (ESOMP); (ii) hub/feeder stations, collecting and distributing the information to/from the other user terminals; and (iii) gateways (GW), which provide the connectivity between the user terminals and the terrestrial infrastructure.

Based on the above definitions, the communication links in the NTN can be classified depending on either the direction or the elements involved in the communication. In particular:

- The *User Link* includes the uplink and downlink communications from/to the user terminals, while the Feeder link includes the uplink and downlink transmissions from/to the hub/feeder or GW.

- The *Forward Link* is the link from the hub/feeder or GW to the user terminals, while the Return Link denotes the link from the user terminals to the hub/feeder or GW. Thus: (i) the Forward Link includes the Feeder (Up)link and the User (Down)link; and (ii) the Return Link includes the User (Up)link and the Feeder (Down)link.
- In addition, the elements in the space/aerial segment might have the capability of communicating with each other. These links are denoted as *Inter-Satellite Links* (ISL).

In Sections 3.2 and 3.3, an introduction to the main characteristics of NTNs is provided in terms of orbits, along with their advantages and disadvantages, constellations and constellation design principles, space mission geometry, and link budget computation.

3.2 Orbits and Constellations

In this section, the basic principles of orbital motion are discussed, including the concept of Keplerian orbits, orbital parameters, and orbital perturbations; these will be exploited to define the different types of orbits, in particular those of interest for NTN, and a few concepts related to the design of NTN constellations.

3.2.1 Principles of Orbital Motion

An *orbit* can be defined as the curved trajectory of an object in space, a *satellite*, around another object, such as that of a planet around the Sun. To a close approximation, the orbits of natural and artificial satellites are described by Kepler's laws of planetary motion: (i) all planets move in elliptical orbits, with the Sun at one focus (in 1609); (ii) the vector from the Sun to a planet sweeps equal areas in equal times (in 1609); and (iii) the square of the period of any planet is proportional to the cube of the semimajor axis of its orbit (in 1618). The mathematical proof was provided a few decades later[1] by Isaac Newton based on the following assumptions: (i) the mass of the satellite, m, is small compared to the mass of the Earth, m_E; (ii) the Earth is spherical and homogeneous; and (iii) the two-body movement occurs in free space, i.e., the satellite and the Earth are the only bodies in the system. In this context, the two main forces governing the motion of the satellite are

_____
1 The proof was published in *Philosophiae Naturalis Principia Mathematica* in 1687, but Newton's studies on orbital motion were actually performed 20 years before as a graduate student.

Newton's Second Law of Motion and the Law of Universal Gravitation; the satellite is said to be in pure orbital motion when these two forces are perfectly balanced. The Law of Universal Gravitation states that any two bodies attract each other with a force proportional to the product of their masses and inversely proportional to the square of the distance, r, between them:

$$\mathbf{F} = -\frac{Gmm_E}{r^3}\mathbf{r} = -\frac{\mu m}{r^3}\mathbf{r} \tag{3.1}$$

where the negative sign accounts for the fact that the gravitational force is directed from the satellite toward the Earth, $G = 6.6743 \cdot 10^{-11}$ m^3/(kg s^2) is the universal gravitational constant, and $\mu \approx Gm_E$ is the Earth's gravitational constant, approximately equals $3.986 \cdot 10^{14}$ m^3/s^2; the actual gravitational constant value is $\mu = G(m + m_E)$, but, since we assumed $m_E \gg m$, we can neglect the satellite's mass. The Second Law of Motion states that the time rate of change of the momentum of a body is equal in magnitude and direction to the net force imposed on it, i.e., $\mathbf{F} = m\ddot{\mathbf{r}}$. Combining the two laws yields to the *orbit equation* (or two-body equation of motion):

$$\ddot{\mathbf{r}} + \frac{\mu}{r^3}\mathbf{r} = 0 \tag{3.2}$$

The above-mentioned formula represents a second-order differential equation governing the motion of a satellite relative to the Earth, for which a solution is given by the *polar equation of a conic section*:

$$r = \frac{a(1 - e^2)}{1 + e \cos \nu} \tag{3.3}$$

where a is the semimajor axis of the elliptical orbit, e is its eccentricity, and ν is the true anomaly (or polar angle), i.e., the angle measured at the Earth's center from the perigee (orbital position corresponding to the minimum value of the distance r) to the satellite, as shown in Figure 3.4. It is worth mentioning that the distance r is measured between the center of the Earth and the satellite, i.e., $r = h_{sat} + R_E$, where h_{sat} denotes the satellite altitude above the ground and R_E the Earth's radius; in the following, under the spherical Earth assumption, $R_E = 6371$ km denotes the mean Earth's radius.

The solution to (3.3) is a conic section, i.e., a curve formed by the intersection of a plane passing through a right angular cone; depending on the inclination of the plane with respect to the cone, different shapes are obtained: circle, ellipse, parabola, or hyperbola. While all conic sections from (3.3) satisfy the orbit equation, as shown in Figure 3.2, only those corresponding to an eccentricity $0 \leq e < 1$ are bounded and, thus, of interest for satellite communications: circular orbits, with $e = 0$, and elliptical orbits, with $0 < e < 1$. When $e > 1$, the satellite escapes the Earth's gravitational pull.

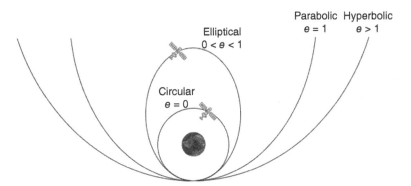

Figure 3.2 Orbit shape as a function of the eccentricity.

Table 3.1 Summary of constants of motion for bounded orbits.

Orbit	Eccentricity	Velocity	Period
Circular	$e = 0$	$v_{sat} = \sqrt{\frac{\mu}{r}}$	$T = 2\pi\sqrt{\frac{r^3}{\mu}}$
Elliptical	$0 < e < 1$	$v_{sat} = \sqrt{\mu\left(\frac{2}{r} - \frac{1}{a}\right)}$	$T = 2\pi\sqrt{\frac{a^3}{\mu}}$

Table 3.1 reports the orbital velocity and the period of a satellite in a bounded orbit, which can be derived based on the laws of conservation of angular momentum and energy [1, 2]. Clearly, for a circular orbit, these constants of motion can be obtained from the elliptical case by setting $a = r$; moreover, it is worth highlighting that the period is independent of the eccentricity, as per Kepler's third law. Figure 3.3 shows the velocity and period of a satellite in a circular orbit as a function of its altitude.

The true anomaly v, the eccentricity e, and the length of the semimajor axis a determine the position of the satellite on the orbital plane. In order to completely define the position of the satellite in space, three additional parameters are needed so as to identify the orbit orientation.[2] Referring to Figure 3.4, the classical (or Keplerian) orbital elements are defined in a Geocentric Celestial Inertial (GCI) coordinate system in which the x-axis is on the equatorial plane and directed toward the vernal equinox,[3] the

_____

2 To find a solution to (2), six constants of integration are needed as the initial conditions.
3 Due to the irregularities in the Earth's rotation, a variation is induced on the x-axis direction. In order to define an inertial coordinate system, a time reference shall be defined for the vernal equinox direction. Typically, January 1, 2000, is used.

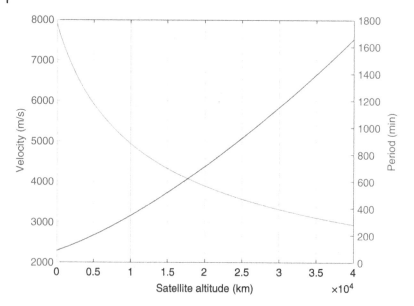

Figure 3.3 Satellite velocity and period in a circular orbit as a function of the altitude.

z-axis is parallel to the North Pole, and the y-axis completes the right-hand system. In this framework, the Keplerian orbital elements are the following:

- Semimajor axis a: It defines the size of the elliptical orbit.
- Eccentricity e: It describes the shape of the elliptical orbit.
- True anomaly v: The angle measured from the perigee to the satellite's position in its orbit.
- Inclination i: The vertical tilt of the orbital plane with respect to the equatorial plane, measured at the ascending node, i.e., the point where the satellite passes through equatorial plane moving from south to north.
- Right Ascension of the Ascending Node (RAAN) Ω: The angle from the vernal equinox to the ascending node, measured as a right-handed rotation around the z-axis.
- Argument of the perigee ω: The angle from the ascending node to the perigee.

It is worthwhile mentioning that the semimajor axis a, the eccentricity e, the RAAN Ω, the argument of the perigee ω, and the inclination i are constant and time-independent parameters, which completely define the orbital plane: its shape (e), size (a), and orbital orientation (i, Ω, ω); the true anomaly v is the only time-dependent term and it determines the current satellite position on the orbit.

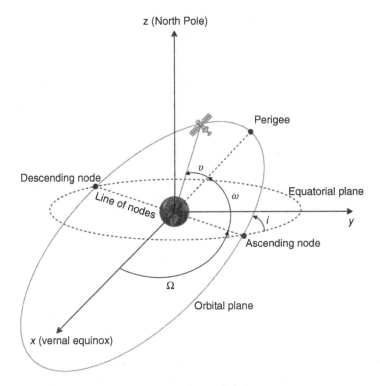

Figure 3.4 Definition of the Keplerian orbital elements.

Depending on the value of the inclination, we can classify the orbit as:

- Direct (or prograde), when $0° \leq i < 90°$, corresponding to a situation in which the satellite orbits the Earth in the same direction of the Earth's rotation.
- Indirect (or retrograde), when $90° < i \leq 180°$, in which the satellite is orbiting the Earth in the opposite direction of the Earth's rotation.
- Polar, when $i = 90°$, i.e., the satellite passes above the Earth's poles.
- Equatorial, when $i = 0°$ (prograde) or $i = 180°$ (retrograde).

Notably, under the assumptions that led to (3.1), the orbit would remain fixed in the introduced GCI frame. However, there are several factors that affect this simple framework and that shall be taken into account in order to properly correct the satellite orbit during its lifetime:

- Third-body interactions: In general, for satellites in low-altitude orbits, the influence of the gravitational attraction of the other planets, the Sun, and the Moon on the satellite is negligible, and its effect is dominated by other perturbations. However, for large orbit altitudes, the Sun and the Moon do impact the orbital parameters and their effect shall be accounted for by means of orbit corrections.

- Non-spherical Earth: The Earth is not spherical as previously assumed, but its shape can be closely approximated as an oblate spheroid with an elliptical equatorial cross-section. The former is particularly detrimental for near-Earth orbits, while the latter causes geostationary satellites to drift in longitude.
- Solar radiation pressure: Its influence is significant for satellites equipped with large solar arrays.
- Atmospheric drag: Satellites in orbits below 1000 km fly through the atmosphere, which causes a loss in their orbital energy and, ultimately, their velocity and altitude. If not properly managed, this effect would lead to the satellite re-entering the Earth's atmosphere and being destroyed.

The above effects shall be taken into account and compensated by means of satellite maneuvering actions, so as to maintain the desired orbit [2, 3].

3.2.2 Types of Orbits

There are various types of orbits, which can be broadly classified based on altitude, orbital plane inclination, and orbit shape and size, as shown in Figure 3.5. The type of service provided by the satellite is the major driver for the definition of its orbit. In general, orbits can be classified as follows [1–3]:

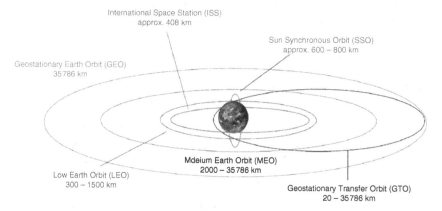

Figure 3.5 Main orbits for artificial satellites around the Earth. Source: Cottin et al. [4]/Springer Nature.

- *Low Earth Orbit (LEO)*, in which the satellite has a low altitude, between 300 and 1500 km, with an inclination that mainly depends on the service to be provided. The possibility to have so many different options for the orbital plane is one of the reasons for which LEO is a widely adopted orbit. For instance, it is used for satellite imaging, since it allows to take high-resolution images, and for the International Space Station (ISS), since it allows easier and less expensive travel for the astronauts. With the start of the New Space Era, thanks to the miniaturization of satellite components and cost reduction for building, launching, and operating smaller satellites, LEO and *very Low Earth Orbit (vLEO)* systems, i.e., satellites below 300 km, have been receiving an ever-increasing attention, as substantiated by the massive industrial endeavors in the race to mega-constellations (e.g. SpaceX Starlink, Amazon Kuiper, OneWeb, Telesat, and LeoSat) and in the development of CubeSats. Due to the significant orbital velocity (approximately 7.78 km/s at 200 km), a large number of satellites (up to thousands or tens of thousands) is required so as to provide global communication services. The most relevant benefits in exploiting this orbit are: (i) the lower over-the-air one-way latency, as low as 0.5 ms at 160 km, which allows very low latency services, a fundamental part of the 5G vertical markets[4]; (ii) the shorter link distances leading to smaller path losses, which allow to either reduce the transmission power for a target performance at the receiver or to provide a larger capacity for a fixed transmission power (this is particularly important for Internet of Things, IoT, services over satellite); (iii) the much narrower beam footprints, allowing to increase the frequency reuse and ultimately the system throughput; (iv) an easier access to spectrum, since Ka and Ku bands can be used on a secondary basis; and (v) the geometry between satellites and User Equipment (UE) varies over time, providing path diversity that can be particularly useful in urban environments. Of course, there are also challenges to be coped with, including: (i) the complexity of the ground segment, in which a network of GWs is needed to track all satellites of the (mega-)constellation (in particular, in the absence of ISL); (ii) the cost of the UE, which is required to track the satellites and to implement fast handover procedures; (iii) the cost of operating and maneuvering the constellation, to accurately

---

4 It is worthwhile mentioning that, over long communication distances (e.g., between London and New York), a fiber optic connection would result in a larger communication delay compared to a LEO constellation with ISLs. In fact, in the fiber, the signal is not traveling at the speed of light, rather at 2/3 of this value, whereas in space, the signal is actually traveling at the speed of light. Thus, despite the larger distance to be traveled, there is an advantage in terms of propagation delay.

keep the phasing among the satellites and avoid collisions; (iv) the need for interference management and coordination techniques, in particular toward geosynchronous systems; and (v) the large initial investment, since to start operating the business a significant portion of the constellation shall be in orbit. Related to the complexity of the ground segment, it is worthwhile mentioning that more advanced payloads with on-board computation capabilities might lead to a reduction of the number of on-ground gateways, by implementing autonomous decision algorithms in space, e.g., for station keeping and collision avoidance. To this aim, ISLs are required, as previously mentioned.

- *Medium Earth Orbit (MEO)*, which comprises a wide range of altitudes, typically between 2000 and 35 786 km. As for LEO and vLEO systems, multiple satellites are needed to provide global services, but the number of satellites is clearly lower. There are two MEO orbits of particular interest: the semi-synchronous and the Molniya orbits. As for the former, it is an MEO orbit at 20 200 km, corresponding to an orbital period of 12 hours; thus, over a day, the satellite passes over the same location twice. This orbit is used for the Global Positioning System (GPS) and for the Galileo system, which actually orbits at 23 222 km. The Molniya orbit was introduced by Russia, with an inclination of 63.4° and $a = 26\,600$ km. It is an orbit with a period of 12 hours, with the advantage of covering high latitudes around the apogee for a large portion of its orbit and with elevation angles close to 90°, i.e., the users are not obstructed by the surrounding environment. This orbit can also be classified as a *High Elliptical Orbit (HEO)*.

- *Geosynchronous Orbit (GSO)*, in which the satellite is precisely at 35 786 km so as to match the Earth's rotation. From an observer on the Earth's surface, the satellite returns to exactly the same location in the sky after an orbital period (equal to a sidereal day). During the orbital period, the satellite will move around this location, with its projection on the Earth typically creating a figure-8 form. When the orbital plane is on the equatorial plane, the satellite is in a *Geostationary Earth Orbit (GEO)* and, apart from orbital perturbations, from an observer's perspective on the Earth, it is in a fixed location in the sky at any time. This orbit has always been used for telecommunication services; in fact, a single GEO satellite covers a large portion of the Earth (approximately, 42.4% of the Earth's surface), requiring just three satellites to provide global coverage. However, by means of simple geometrical considerations, it can be shown that GEO satellites, being on the equatorial plane, are not visible at latitudes above 81.3°. Another advantage is that, being at a fixed location in the sky, the UE antenna can point in a fixed direction.

- *Polar Orbit (PO)*, in which the satellite passes relatively close to the two poles. Typically, a deviation from the pole axis up to 20° is considered acceptable to still classify the orbit as polar. They are a special case of LEO, since the orbit is typically below 1000 km.
- *Sun-Synchronous Orbit (SSO)* is a special case of PO, in which the satellites are traveling above the poles and are also synchronous with the Sun, i.e., they pass over the same location at the same local mean solar time. It is particularly used to observe the same location at the same time of the day over long periods for scientific reasons.
- *Geostationary Transfer Orbits (GTO)* are special orbits that are used to place the GEO satellite in its correct position. More specifically, instead of launching a GEO satellite directly to the correct orbit, which can be costly, it is typically launched to a lower altitude, and then, through the GTO, it is placed in its orbital slot.
- *Lagrange points* are over a million km from Earth and, actually, they do not orbit the Earth in the strict sense. An L-point is a specific location in far space, where the gravitational attraction from the Sun and the Earth is perfectly balanced; as a consequence, the satellite is in a stable position compared to the Earth. This orbit is used for space-based observatories and telescopes, such as the recently launched James Webb telescope that will reach the second Lagrange point (1.5 million km) during the first half of 2022. With respect to the Lagrange points of the Earth-Moon system, it is worthwhile mentioning that points L4 and L5 are located at approximately 400 000 km from the Earth, and they are particularly important due to the better stability that is offered compared to L1, L2, and L3, i.e., a reduced need for station-keeping fuel reserves.

Apart from GEO satellites, which appear to be at a fixed location in the sky, the area in which a satellite can provide a service, denoted as *individual satellite coverage*, moves accordingly with the satellite's movement in its orbit. Referring to Figure 3.6, this leads to a very important distinction in terms of system coverage: (i) the *instantaneous system coverage*, which can be defined as the aggregation at a given time of the coverage areas of the individual satellites in the constellation; and (ii) the long-term system coverage, which is the aggregation of the instantaneous system coverage areas over time, i.e., the overall area that is covered by all of the satellites during their orbital revolutions. Clearly, the coverage area should always include the service area, i.e., the on-ground area in which the service is to be provided. For real-time services, the user should be able to communicate at any moment and, thus, the service area should be always located within the instantaneous system coverage; for non-real-time services, it is

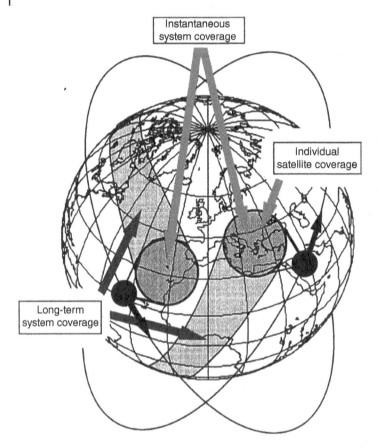

Figure 3.6 Instantaneous and long-term system coverage. Source: Maral and Bousquet [5]/John Wiley & Sons.

sufficient that with a given periodicity the user is allowed to communicate and, thus, the service area should be located within the long-term system coverage [5].

As discussed below, in the context of New Radio (NR) communications via satellite, not all of the orbits introduced above are of interest to NTN. In particular, the orbits currently being considered of interest by 3rd Generation Partnership Program (3GPP) for non-terrestrial systems are GEO and LEO, with implicit support of MEO and HAPS systems, consisting of an airborne vehicle equipped with the NTN payload placed at an altitude between 8 and 50 km.

Finally, it shall be noticed that the orbits discussed above are the main categories that are identified in Satellite Communications. Notably, each orbit has many potential implementations, depending on the specific design and optimization of the orbital parameters; moreover, many possibilities are also available when designing a new constellation by including elements from multiple orbits into an hybrid solution. For instance, the Phase 1 of the Starlink constellation includes 5 orbital shells with different inclinations and number of satellites:

- Shell 1: 1548 satellites with 53.0° inclination at 550 km;
- Shell 2: 720 satellites with 70° inclination at 570 km;
- Shell 3: 348 satellites with 97.6° inclination (allowing to cover the polar regions) at 560 km;
- Shell 4: 1548 satellites with 53.2° inclination at 540 km;
- Shell 5: 172 satellites with 97.6° inclination (again polar regions are covered) at 560 km.

3.2.3 Constellation Design

As discussed above, a single satellite covers a limited portion of the Earth's surface and multiple satellites are thus needed to provide global services. A *constellation* can be defined as a number of similar satellites, of a similar type and function, designed to be in similar, complementary orbit for a shared purpose and under a shared control. The design of a satellite constellation is a key aspect, since it is driven by the type of service and extent of the coverage to be provided, and it defines the orbital parameters of each satellite. There are many possible approaches to design a constellation. In the framework of 3GPP NTN activities, the attention is currently oriented toward circular orbits, i.e., $r = a$. A very popular category of circular orbit geometries is the *Walker Constellation* [6, 7], which is also proposed as a reference in 3GPP TR 38.821 [8].

A Walker constellation is based on a simple yet effective design strategy for distributing the satellites in the constellation, in which all satellites are placed on circular orbits at the same altitude; the constellation is identified by means of four parameters: (i) the *inclination* of the orbital planes (all planes have the same inclination), i; (ii) the *total number of satellites* in the constellation, t; (iii) the *number of orbital planes*, p; and (iv) the *inter-plane phasing*, i.e., the relative spacing between satellites in adjacent orbital planes, f. The compact notation to completely characterize a Walker constellation is $i : t/p/f$; clearly, the number of satellites per orbital plane can be obtained as $S = t/p$. There are two main variants for a Walker constellation, which differ in how the ascending nodes between the orbital planes are distributed, as also shown in Figure 3.7:

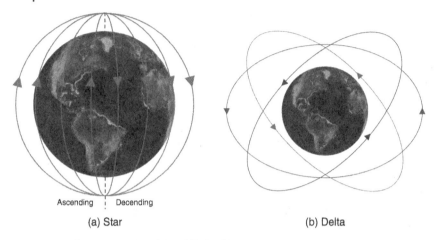

Ascending ¦ Decending

(a) Star   (b) Delta

Figure 3.7 Walker Star (a) and Delta (b) constellations.

- Walker Star: All of the orbital planes cross over the poles or near them (if viewed from the point of intersection, the orbital planes form a star). By geometric construction, the ascending nodes of the constellation are approximately evenly spaced over 180°, since after crossing the pole, the satellites will move in the opposite direction. This spacing is different between the contra-rotating planes (the two central planes in Figure 3.7a), since the satellites are moving in opposite directions.[5] This aspect leads to a technical challenge in case ISLs are implemented; in fact, for satellites belonging to the adjacent contra-rotating planes the relative orbital speed of one of them with respect to the other is twice the orbital speed of a single satellite. This makes the ISL quite challenging to be established, if possible at all. For the contra-rotating planes, the separation is also slightly reduced compared to that between co-rotating planes in order to ensure a full and overlapping ground coverage. Finally, it is worthwhile noticing that the larger the altitude, the closer are the satellites to adjacent orbital planes; this aspect might pose some issues in terms of interference management and in satellite handover, which might be triggered very often. This constellation was used, for instance, in the Iridium system.
- Walker Delta: This is a more general case, with the orbital planes inclined by a certain angle i and the ascending nodes being evenly spaced over 360°.

---

5 The orbital planes defined in the range from 0° to 180° have satellites with ascending movement. Once the satellites cross the pole, they will have a descending movement and, thus, by geometry construction of the constellation there will always be counter-rotating planes.

Such even spacing leads to ascending and descending planes that continuously overlap, instead of being separated as in the star configuration. Typically, no coverage is provided at the poles, given the common inclination of the orbital planes, and the best diversity is obtained at mid-latitudes. This constellation is used for the Galileo system with a $56° : 24/3/1$ configuration.

In the context of satellite constellations, it is worthwhile mentioning that for a larger number of satellites in orbit, the elevation angles at which the potential serving satellites are seen by the users becomes closer to 90°. This is particularly important as for higher elevation angles, the impact of potential obstacles and multi-path fading is more limited. Moreover, larger elevation angles also lead to an easier coexistence with terminals connecting to other NTN systems or with terrestrial networks. In particular, with a larger number of satellites providing coverage at larger elevation angles, it is easier to meet the requirements on the interference protection ratios, in addition to other technical and electronic enhancements such as the use of active antennas allowing adapting beam shaping and user-centric beamforming solutions.

3.2.4 Satellite Orbit Determination and Prediction

The determination of the satellite orbit consists of the estimation of the most likely orbit that the satellite followed based on previous noisy measurements of its position and velocity. This task can be performed based on the following steps, as shown in Figure 3.8 [9]:

1. The satellite position and velocity vectors are estimated and associated with a timestamp, typically based on Global Navigation Satellite System (GNSS) on-board measurements. Notably, the accuracy of the GNSS measurements is impacted by the GNSS receiver characteristics, the satellite orbit, and the number of the GNSS satellites in view.
2. The above measurements are collected at the NCC, where the actual orbit determination (OD) is performed. Typically, the GNSS measurements are provided to the ground segment through telemetry channels. The actual reporting period for such measurements is implementation-specific; however, it is directly related to the desired accuracy.
3. The OD is performed, based on a more or less complex approach depending on the implemented model, the amount of GNSS measurements, and the used algorithm. Usually, two filtering techniques are used: Kalman filtering or least-square filtering. The latter provides more accurate results but is more computationally demanding.

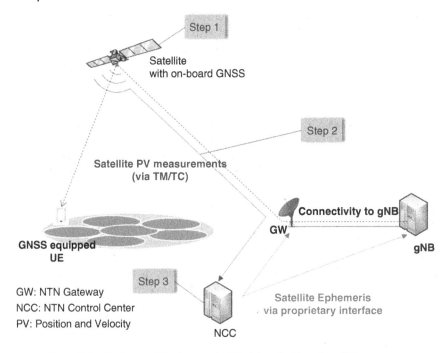

Figure 3.8 System architecture and orbit determination steps [9].

The best precision in OD is obtained for satellite systems designed for navigation, which deliver accurate GNSS measurements with a position Root Mean Square Error (RMSE) of 0.05 m and a velocity RMSE of 0.05 mm/s. However, commercial telecommunication satellites cannot meet these performances due to the large cost and, moreover, such high precision is not required. Typical OD systems provide an RMSE that equals 0.5 m and 0.5 mm/s for the position and velocity. Finally, in low-cost systems, the precision is again worse by an order of magnitude, leading to 5 m and 5 mm/s RMSE values.

Notably, the OD results can be used in orbit prediction (OP), i.e., to derive updated satellite ephemeris, which can be used to predict future satellite position and velocity vectors. Such information can be useful for, e.g., the determination of when the satellite will be above an area of interest, anticipating collision avoidance operations, anticipating future feeder link switch-over operations, and performing Doppler compensation. Clearly, the accuracy of the OD that is needed for these operations depends on the exact application. Finally, the NCC can share the ephemeris with other network elements that are responsible for the operations, such as the NTN GW. The

OP accuracy can depend on several factors, including: (i) the accuracy of the OD used to derive the ephemeris, as mentioned above; (ii) the accuracy of the orbit propagation model; and (iii) the time horizon over which the OP is performed.

3.3 Propagation and Link Performance

In space mission geometry, the objectives are often to: (i) understand the apparent position and motion of objects on the Earth's curved surface as seen from the satellite; and (ii) know the satellite's position and trajectory as seen from a given location on the Earth's curved surface. This type of analysis requires some basic concepts on Keplerian, discussed in Section 3.2.1, and, typically, principles on direction-only and celestial sphere geometry. In this section, we exploit space mission geometry to characterize Doppler and propagation delays and to define the individual and overall link budgets.

3.3.1 Earth–Satellite Geometry

For a satellite in a given orbital position at a distance $r = h_{sat} + R_E$ from the Earth's center and a UE at a location on the Earth's surface within the satellite visibility, several geometrical parameters can be defined, as shown in Figure 3.9:

- The *elevation angle* ε, defined as the angle between the tangent to the Earth's surface at the UE location and the satellite direction from the UE location. By simple geometry considerations, this angle is always between 0° and 180°.
- The *angular Field of View (FoV)* ρ, measured at the satellite between the Earth's center direction and the geometrical horizon on the Earth's surface.
- The *nadir angle* η, defined as the angle measured at the satellite from the Earth's center direction to the UE direction, with $0° \leq \eta \leq \rho$.
- The *slant range d*, which is the distance between the UE and the satellite.
- The *Earth central angle* λ_0, computed as the angle at the Earth's center from the satellite direction to the direction of the geometrical horizon. This angle measured between the satellite and the UE location is the *Earth central angle at the UE*, λ_{UT}.

The angular FoV provides a measure of the portion of Earth's surface that is visible from the satellite, and it can be computed as follows:

$$\sin \rho = \cos \lambda_0 = \frac{R_E}{R_E + h_{sat}} \tag{3.4}$$

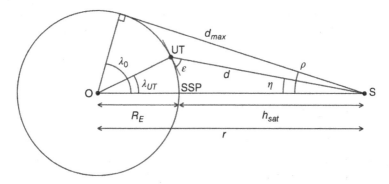

Figure 3.9 Reference diagram for the Earth–satellite geometry.

Clearly, the angular FoV increases for increasing altitudes. From the above equation, it is also possible to obtain the instantaneous access area[6] (IAA), i.e., the portion of the Earth's surface that falls within the satellite FoV:

$$IAA = K_{IAA}(1 - \cos \lambda_0) \tag{3.5}$$

where K_{IAA} is a constant defined by the units in which such area is desired; for instance, to compute (3.5) in km², $K_{IAA} = 2.556\,041\,87 \cdot 10^8$ km², while for steradians $K_{IAA} = 2\pi$. It can be easily shown that the IAA is always upper bounded by one-half of the Earth's surface, since $\lambda_0 \to \pi/2 \Longrightarrow IAA \to 2\pi$. For a GEO satellite, $h_{sat} = 35\,786$ km leads to $IAA \approx 5.3336$ steradians, i.e., approximately 42.4% of the Earth's surface as discussed above.

Referring to Figure 3.9, the Sub Satellite Point (SSP), i.e., the projection of the satellite position on the Earth's surface, is the location on the Earth's surface that is at minimum slant range, maximum elevation angle, and minimum Earth central angle for a location in the satellite FoV:

$$d_{SSP} = h_{sat} \Longrightarrow r_{SSP} = h_{sat} + R_E \tag{3.6}$$

$$\varepsilon_{SSP} = 90° \tag{3.7}$$

$$\lambda_{SSP} = 0° \tag{3.8}$$

---

6 It shall be noticed that the instantaneous access area is not the instantaneous system coverage previously defined. The former is purely defined by geometry, while the latter is defined based on the multiple or single beam coverage provided by the satellite antenna(s).

On the other hand, focusing on the horizon, we obtain the maximum slant range, minimum elevation angle, and maximum Earth central angle values for a location in the satellite FoV:

$$d_{hz} = \sqrt{(h_{sat} + R_E)^2 - R_E{}^2} = R_E \tan \lambda_0 \tag{3.9}$$

$$\varepsilon_{hz} = 0° \tag{3.10}$$

$$\lambda_{hz} = \lambda_0 \tag{3.11}$$

For a generic location in the FoV, the computation of the geometrical parameters is typically performed by observing that ε, λ_{UT}, and η satisfy the following condition:

$$\varepsilon + \lambda_{UT} + \eta = 90° \tag{3.12}$$

In general, one of these angles is known and the others can be computed by exploiting the above equation and the knowledge of the angular FoV of the satellite from (3.4):

• If λ_{UT} is known, we can compute η by observing that

$$\tan \eta = \frac{\sin \rho \sin \lambda_{UT}}{1 - \sin \rho \sin \lambda_{UT}} \tag{3.13}$$

• If η is known, then ε can be computed as

$$\cos \varepsilon = \frac{\sin \eta}{\sin \rho} \tag{3.14}$$

• If ε is known, then

$$\sin \eta = \cos \varepsilon \sin \rho \tag{3.15}$$

Once the elevation angle at the UE location has been computed, the slant range is given by:

$$d = \sqrt{(R_E + h_{sat})^2 - R_E{}^2 \cos^2 \varepsilon} - R_E \sin \varepsilon \tag{3.16}$$

Figure 3.10 shows the slant range as a function of the elevation angle ε for different values of the satellite altitude h_{sat}. It can be noticed that the minimum is always obtained at $\varepsilon = 90°$.

To conclude the discussion on the basic principles of Earth–Satellite geometry, it is worthwhile highlighting that Satellite Communication systems are never designed so as to cover users at elevation angles down to 0°. In fact, for a satellite located exactly at the horizon from the user's perspective, a foreshortening effect arises that can distort the view. In addition, and more importantly, obstacles are much more likely to obstruct the Line-Of-Sight (LOS) between the UE and the satellite, leading to Non-LOS

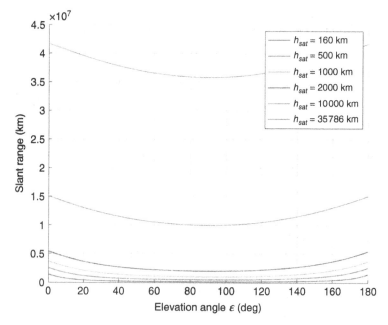

Figure 3.10 Slant range as a function of the elevation angle for different satellite altitudes.

(NLOS) conditions that are detrimental to the link performance. Thus, there is always a minimum elevation angle that defines the on-ground coverage requirement, which leads to a reduced angular FoV expressed in terms of a maximum nadir angle.[7] For a minimum elevation angle ε_{min}:

$$\sin \eta_{max} = \frac{R_E}{R_E + h_{sat}} \cos \varepsilon_{min} \implies \eta_{max} = \sin^{-1} \left(\frac{R_E}{R_E + h_{sat}} \cos \varepsilon_{min} \right)$$
$$(3.17)$$

Figures 3.11 and 3.12 show an example of FoV for a GEO and a LEO satellite, respectively, located on the equatorial plane and with SSP at $Lat = 0°, Lon = 0°$.

3.3.1.1 Delay Characterization

In the context of NR via satellite, different figures of merit are required to characterize the propagation delay [10]. The *one-way propagation delay*, τ,

---

7 This requirement can also be expressed in terms of a maximum angular FoV. However, since the FoV is a geometrical property defined by the satellite altitude, it is formally more correct to associate the constraint to the maximum nadir angle at which users are allowed.

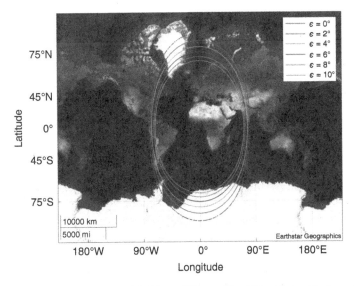

Figure 3.11 Example of FoV for a GEO satellite, SSP at $Lat = 0°$, $Lon = 0°$.

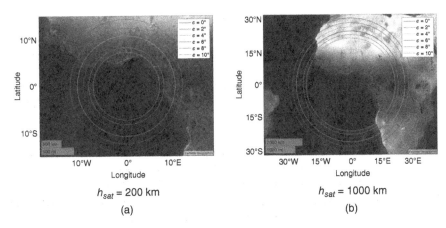

$h_{sat} = 200$ km

(a)

$h_{sat} = 1000$ km

(b)

Figure 3.12 Example of FoV for a LEO satellite at 200 km (a) and 1000 km (b) altitude, SSP at $Lat = 0°$, $Lon = 0°$.

is the over-the-air latency due to propagation. For transparent payloads, the satellite basically acts as a Radio Frequency (RF) repeater and, thus, the over-the-air latency shall be computed between the GW and the user:

$$\tau_T = \frac{d_{GW-sat} + d_{sat-UE}}{c} \quad (3.18)$$

where c is the speed of light, d_{GW-sat} the slant range between the GW and the satellite, and d_{sat-UE} that between the satellite and the UE. For a regenerative payload, the connection is established between the UE and the satellite and, thus:

$$\tau_R = \frac{d_{sat-UE}}{c} \qquad (3.19)$$

It is worth to be mentioned that, with regenerative payloads, the one-way delay might also be computed between the UE and the GW. In fact, the network entity with which the user terminal is communicating depends on the type of functional split that is implemented, i.e., on which layers are implemented on-board and which ones on-ground, as discussed in Chapter 4. However, the case in which the intended layer is implemented on-ground is covered by Equation (3.18) and, thus, in the following, we assume that for a regenerative payload the communication is always between the UE and the satellite. For the characterization of the propagation delay, also the *Round Trip Time* (*RTT*) is an important parameter, which is simply defined as twice the one-way delay:

$$RTT_i = 2\tau_i \qquad (3.20)$$

where $i = T, R$ depending on the payload type. In the context of 3GPP NTN scenarios, for the propagation delay analyses, it is assumed that the UE and the GW minimum elevation angles, leading to the maximum delays, are given by $\varepsilon_{UE} = 10°$ and $\varepsilon_{GW} = 10°$, respectively [8]. From these values, Equation (3.16) can be used to obtain the slant ranges and, thus, the latency. Figure 3.13 shows the minimum (user at the SSP, $\varepsilon_{UE} = 90°$) and maximum (user at $\varepsilon_{UE} = 10°$) one-way propagation delay for transparent and regenerative payloads. Table 3.2 reports the specific slant ranges, one-way delay, and RTT for GEO and LEO satellites, with $h_{sat} = 600$ km and $h_{sat} = 1200$ km (two configurations of interest for 3GPP NTN). The advantage of terminating the protocols on-board is clear, with latencies basically halved.

Notably, there are NR procedures, in particular at physical (PHY) and Medium Access Control (MAC) layer, with strict requirements in terms of latency in the Control Plane (CP), e.g., Random Access (RA) and Timing Advance (TA) [8, 11–13]. In this context, compared to legacy terrestrial networks, the significantly larger delays pose a technical challenge on the path to the integration of the NR Air Interface and its procedures in the NTN context. While specific aspects related to the various protocols will be addressed in the next chapter, here we only focus on the computation of the maximum differential delay, i.e., the difference in the propagation delay between the two positions in a given beam that corresponds to the minimum and maximum slant ranges, i.e., minimum and maximum delay. In [8], it is

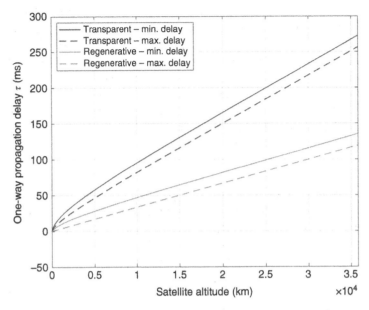

Figure 3.13 Minimum and maximum one-way propagation delay for transparent and regenerative payloads.

Table 3.2 Summary of one-way and RTT delays for GEO and LEO satellites at 600 and 1200 km, with transparent and regenerative payloads.

| Scenario | Path | Transparent payload | | Regenerative payload | | |
		Slant range (km)	Latency (ms)	Path	Slant range (km)	Latency (ms)
GEO	GW-sat	41 121	137.17	sat-UE	40 581	135.36
	sat-UE	40 581	135.36			
	one-way		272.53	one-way		135.36
	RTT		545.06	RTT		270.72
LEO, $h_{sat} =$ 600 km	GW-sat	2328	7.77	sat-UE	1931	6.44
	sat-UE	1931	6.44			
	one-way		14.21	one-way		6.44
	RTT		28.42	RTT		12.88
LEO, $h_{sat} =$ 1200 km	GW-sat	3572	11.91	sat-UE	3131	10.44
	sat-UE	3131	10.44			
	one-way		22.35	one-way		10.44
	RTT		44.7	RTT		20.88

clearly stated that the maximum differential delay shall be computed with respect to the maximum beam footprint, irrespective of the elevation angle; clearly, this definition does not preclude the possibility to have narrower beams in the coverage, but it only defines the maximum value that shall be supported. The maximum beam footprint is set to be 3500 and 1000 km for GEO and LEO systems, respectively; this leads to $\Delta\tau = 10.3$ ms for GEO and $\Delta\tau = 3.12$ ms or $\Delta\tau = 3.18$ ms for LEO satellites at 600 or 1200 km. These are the values to be taken into account when evaluating the different timers that are present in the PHY and MAC procedures.

3.3.1.2 Doppler Characterization

For the Doppler shift, we shall discern between GEO and LEO systems, due to the very different orbital motions. As for the latter, the Doppler shift is maximum when the UE is located on the ground track, i.e., on the projection of the orbital plane on the Earth's surface. In this case, a simplified formula for the computation of the Doppler shift is proposed in [14]:

$$f_d(t) = \frac{f_0}{c}\omega_{sat}R_E \cos\varepsilon(t) \qquad (3.21)$$

where:

$$\omega_{sat} = \sqrt{\frac{\mu}{(R_E + h_{sat})^2}} \qquad (3.22)$$

is the angular velocity of the satellite, f_0 is the carrier nominal frequency, and $\varepsilon(t)$ is the elevation angle as a function of time. Equation (3.21) is obtained assuming a fixed on-ground receiver; in case the receiver is moving, then the angular speed ω_{sat} shall be modified so as to take into account the user's speed, i.e., it should be a *net angular speed*. While for GEO (as well as HAPS) the main contribution to the overall Doppler is that related to the users' movement, in LEO systems the most impacting factor is, clearly, the satellite's large orbital speed. Finally, it is worthwhile mentioning that, similarly to the delay analysis, several procedures in the system might be affected by the differential Doppler shift. In particular, for Orthogonal Frequency Division Multiplexing (OFDM)-based waveforms, this value should not exceed a few percentages of the sub-carrier spacing (SCS) (Table 3.3) [12–14].

For GEO systems, as previously mentioned, in the absence of any perturbation, the satellite is perfectly fixed in space, and, thus, the only relative velocity vector between the satellite and an on-ground user would be related to the movement of the latter. However, the GEO satellite is not perfectly fixed, since it slightly moves around its correct orbital location, due to the orbital perturbations that we discussed in the previous section. The on-board maneuvering systems shall keep the satellite inside a limited box, as shown

Table 3.3 Doppler characterization for LEO satellites.

Scenario	Carrier frequency (GHz)	Max. Doppler shift (kHz)	Relative Doppler	Max. Doppler shift variation
LEO h_{sat} = 600 km	2	±48	0.0024%	−544 Hz/s
	20	±480	0.0024%	−5.44 kHz/s
	30	±720	0.0024%	−8.16 kHz/s
LEO h_{sat} = 1500 km	2	±40	0.002%	−180 Hz/s
	20	±400	0.002%	−1.8 kHz/s
	30	±600	0.002%	−2.7 kHz/s

Source: Adapted from [10].

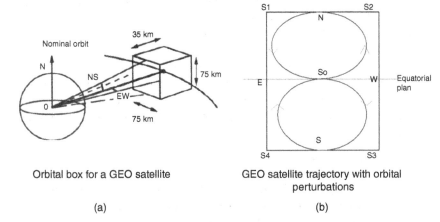

Orbital box for a GEO satellite

(a)

GEO satellite trajectory with orbital perturbations

(b)

Figure 3.14 GEO satellite orbital box (a) and trajectory with perturbations (b) [10].

in Figure 3.14a; inside this box, the satellite usually follows a figure-8 trajectory over a 24-hours period, as depicted in Figure 3.14b.

In TR 38.821, some examples are provided assuming a satellite located at 10°E on the equatorial plane and considering both a high-speed train (500 km/h) and a plane (1000 km/h). Table 3.4 reports the Doppler shift values obtained assuming a fixed satellite with moving users and a moving satellite, along the trajectory in Figure 3.14b, with fixed users for different carrier frequencies. These values are relatively low and compatible with the OFDM structure of the NR waveform.

3.3.2 Link Performance

In order for a user terminal to receive (send) information from (to) the GW, a satellite link includes at least one uplink and one downlink and, if

Table 3.4 Doppler characterization for GEO satellites.

Scenario	Doppler shift (Hz)		
	2 GHz	20 GHz	30 GHz
Fixed GEO, train in the opposite direction	−707	−7 074	−10 612
Fixed GEO, plane in the opposite direction	−1414	−14 149	−21 224
GEO moving from S2 to S1	−0.25	−2.4	−4.0
GEO moving from S1 to S4	2.25	22.5	34

Source: Adapted from [10].

present, one or more ISLs. The performance on the overall end-to-end link is obtained as a combination of that of the individual links. A fundamental step in the design of any NTN system is the computation of the *link budget*, i.e., the budget of all power gains and losses over the communication link.

3.3.2.1 Antenna Parameters and Received Power

An isotropic transmitter radiates its available power P_T uniformly in all directions; in free space, this power, at a certain distance d, is uniformly distributed over a sphere, leading to the following *power spectral density* (*PSD*):

$$\Phi = \frac{P_T}{4\pi d^2} \ (\text{W/m}^2) \tag{3.23}$$

In practical systems, directional antennas are implemented, characterized by an antenna gain in a given transmission direction (ϑ_T, φ_T), $G(\vartheta_T, \varphi_T)$, defined as the ratio between the power radiated per unit solid angle by the antenna in the (ϑ_T, φ_T) direction, $P(\vartheta_T, \varphi_T)$, to the power radiated per unit solid angle by an isotropic source:

$$G(\vartheta_T, \varphi_T) = \frac{P(\vartheta_T, \varphi_T)}{P_T/4\pi} \tag{3.24}$$

The direction $(\vartheta_T, \varphi_T) = (0, 0)$ is called the *antenna boresight* and it corresponds to the direction of the maximum antenna gain, G_{tx}. In this direction, the PSD at distance d becomes:

$$\Phi = \frac{P_T G_{tx}}{4\pi d^2} = \frac{EIRP}{4\pi d^2} \ (\text{W/m}^2) \tag{3.25}$$

where $EIRP = P_T G_{tx}$ is the Effective Isotropic Radiated Power (EIRP) and it provides a measure of the total emitted power on the downlink. It is worthwhile highlighting that, for the *reciprocity theorem*, the antenna equipment has the same operating behavior for both transmission and reception.

At the receiver side, the power captured by an antenna equipment with effective A_{eff} is:

$$P_R = \Phi A_{eff} = \frac{EIRP}{4\pi d^2} A_{eff} \ (\text{W}) \tag{3.26}$$

Notably, the effective area of an antenna can be written as a function of the maximum receiving antenna gain, G_{rx}:

$$A_{eff} = G_{rx} \frac{\lambda^2}{4\pi} \tag{3.27}$$

where $\lambda = c/f$ is the signal wavelength. Including (3.27) in (3.26), we obtain the following link budget equation:

$$P_R = \frac{EIRP}{4\pi d^2} G_{rx} \frac{\lambda^2}{4\pi} = P_T G_{tx} G_{rx} \left(\frac{\lambda}{4\pi d}\right)^2 = \frac{P_T G_{tx} G_{rx}}{L_{FS}} \ (\text{W}) \tag{3.28}$$

where $L_{FS} = \left(\frac{4\pi d}{\lambda}\right)^2$ is the free space loss.

3.3.2.2 Additional Losses
In practical systems, there are many factors that negatively impact the link budget in (3.28), which shall be taken into account to correctly design the system. The most relevant are:

- Antenna depointing and feeder losses: As shown in Figure 3.15, directional antennas are used and they are usually not ideally aligned. Moreover, the performance is evaluated as a function of the power emitted by the transmitting amplifier, $P_{tx} = P_T L_{F,\,tx}$, and the power at the receiver input, $P_{rx} = P_R/L_{F,\,rx}$, leading to:

$$P_{rx} = \frac{P_{tx} G_{tx}(\vartheta_T, \varphi_T) G_{rx}(\vartheta_R, \varphi_R)}{L_{FS} L_{F,tx} L_{F,rx}} \ (\text{W}) \tag{3.29}$$

where $L_{F,\,tx}$ and $L_{F,\,rx}$ denote the feeder losses at the transmitter and the receiver, respectively.

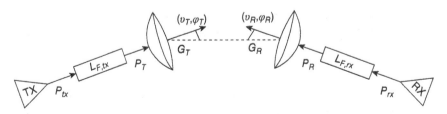

Figure 3.15 Link configuration.

- Depolarization losses: The polarization of the receiving antenna might be not the same as the transmitted signal. The two main factors leading to such misalignment are a mismatch in the orientation of the receiving antenna and the propagation through the atmosphere, introducing a depolarization loss L_{pol}.

- Atmospheric losses: Communication links involving NTN elements are deeply affected by propagation through the troposphere (from ground to 15 km) and the ionosphere (between 70 and 1000 km). ITU-R Recommendation P.618 provides a detailed overview of the procedures and the list of additional recommendations that are needed to compute the corresponding losses [15]. Relative to the free space loss, the following contributions are the most significant, in particular for frequencies above 10 GHz: (i) gaseous absorption, $L_{gas}(p)$; (ii) attenuation due to rain, $L_{rain}(p)$; (iii) attenuation due to clouds, $L_{cloud}(p)$; and (iv) scintillation, $L_{scint}(p)$. These losses have different characteristics, but in general, they all depend on the carrier frequency, the geographic location of the UE, and the elevation angle. Their values are expressed as a function of the probability of the atmospheric event, p; notably, the larger such probability, the more likely to happen is the atmospheric event and, thus, the lower the loss. For a given probability, the corresponding link budget reports values of the received power that are not exceeded for a portion p of the average year, i.e., the service availability on an average year is $1 - p$. It shall be mentioned that there are also other factors that might arise: focusing and defocusing, e.g., decrease in the antenna gain due to wave-front incoherence and attenuation due to sand and dust storms; however, for elevation angles above 10°, which is typically the case as discussed above, they can be neglected. The overall attenuation due to atmospheric phenomena is [15]:

$$L_{atm}(p) = L_{gas}(p) + \sqrt{(L_{cloud}(p) + L_{rain}(p))^2 + L_{scint}^2(p)} \qquad (3.30)$$

- Large-scale effects: Depending on the satellite elevation angle, the link has a certain probability to be in NLOS conditions, i.e., to be obstructed by buildings, foliage, or other obstacles; even for a link in LOS conditions, slow fading effects can arise, leading to the need for including a shadowing margin in the path loss computation. The procedure to compute these effects is reported in 3GPP TR 38.811 [10], leading to the following shadow margin loss:

$$L_{SM} = SF + CL \qquad (3.31)$$

where $SF \sim \mathcal{N}\left(0, \sigma_{SF}^2\right)$ is a normal random variable modeling the *slow fading* on a logarithmic scale, σ_{SF}^2 is its variance, and CL is the *clutter loss*, i.e., the loss due to obstructions in NLOS conditions. These variables depend on the propagation environment, which can be urban, dense urban, or rural, and on the elevation angle. The detailed steps and parameters to compute (3.31) are provided in [10].

This equation provides overall path loss to be used in the link budget Equation (3.29):

$$P_{rx} = \frac{P_{tx} G_{tx}(\vartheta_T, \varphi_T) G_{rx}(\vartheta_R, \varphi_R)}{L_{FS} L_{pol} L_{F,tx} L_{F,rx} L_{atm} L_{SM} L_{ad}} \text{ (W)} \tag{3.32}$$

where L_{ad} is introduced to include any additional loss not covered by those discussed above and, for the sake of clarity, the probability of atmospheric event p is not explicitly reported.

3.3.2.3 Individual Link Performance

NTN links are typically evaluated in terms of the C/N_0 or C/N, i.e., the ratio between the received power at the receiver input and the noise PSD, N_0, or noise power, N, respectively. Notably, assuming an equivalent noise bandwidth B, an antenna noise temperature at the receiver T_a, and a receiver with noise figure N_f:

$$N_0 \text{ (dB)} = N_f \text{ (dB)} + 10\log_{10}\left(T_0 \text{ (K)} + (T_a \text{ (K)} - T_0 \text{ (K)})10^{-0.1 N_f \text{ (dB)}}\right)$$
$$+ \log_{10}\kappa \text{ (dBW/K/Hz)} \tag{3.33}$$

where T_0 is the reference ambient temperature and κ the Boltzmann constant. In clear-sky conditions, i.e., without any rain formation, the antenna temperature T_a on the downlink (Space-to-Earth) is computed as [5]:

$$T_a = T_{sky} + T_{ground} \tag{3.34}$$

with T_{sky} being the sky noise temperature, typically approximated with the brightness temperature as a function of the elevation angle, and T_{ground} representing the noise collected from the ground through the sidelobes. When there is a rain formation, it acts as an attenuator on the signal and, thus, the antenna temperature on the downlink becomes:

$$T_a = \frac{T_{sky}}{L_{rain}} + T_m\left(1 - \frac{1}{L_{rain}}\right) + T_{ground} \tag{3.35}$$

This effect can be neglected on the uplink, since the antenna on-board the satellite is mainly looking at the Earth, which has a much larger noise temperature; in general, a conservative value equal to 290 K is considered.

Merging equations from (3.29) to (3.35), the individual link performance can be written as:

$$\frac{C}{N_0} = \frac{P_{rx}}{N_0} = EIRP \cdot \frac{1}{L} \cdot \frac{G}{T} \cdot \frac{1}{\kappa} \tag{3.36}$$

where: (i) $EIRP = \frac{P_{tx}G_{tx}(\vartheta_T,\varphi_T)}{L_{F,tx}}$ completely characterizes the transmitting equipment; (ii) $L = L_{FS}L_{atm}L_{SM}L_{ad}$ is the overall propagation loss; and (iii) $\frac{G}{T} = \frac{G_{rx}(\vartheta_R,\varphi_R)}{L_{F,rx}L_{pol}T_a}$ is the receiver figure of merit, which provides a measure of the receiver sensitivity. Clearly, from (3.36), we obtain the following Carrier-to-Noise Ratio (CNR):

$$\frac{C}{N} = EIRP \cdot \frac{1}{L} \cdot \frac{G}{T} \cdot \frac{1}{\kappa B} \qquad (3.37)$$

3.3.2.4 Overall Link Performance

Based on the individual link performance in (3.35) and (3.36), it is possible to compute the overall link performance from one ground station to another, which involves (without ISLs) an uplink from the transmitting ground station to the satellite and a downlink from the satellite to the receiving station. To this aim, let us denote as G the total gain between the satellite receiver input and the receiving ground station input, i.e., $G = G_{SR}G_{tx}G_{rx}/L_{F,\,tx}L_{F,\,rx}L_D$, with G_{SR} being the satellite repeater gain and L_D the downlink path loss. The total noise spectral density at the receiver station is given by the sum of the downlink noise spectral density and that on the uplink re-transmitted by the satellite:

$$(N_0)_T = (N_0)_D + G(N_0)_U \qquad (3.38)$$

Thus, from (3.35), it can be easily shown that [5]:

$$\left(\frac{C}{N_0}\right)_T^{-1} = \left(\frac{C}{N_0}\right)_D^{-1} + \left(\frac{C}{N_0}\right)_U^{-1} \qquad (3.39)$$

In the overall link performance for NTN systems, also inter-modulation effects arising from nonlinearities in the on-board amplifiers and the total interference shall be included. The latter can be originated from other satellites or other Earth stations interfering with the intended terrestrial terminals. Taking these factors into account leads to:

$$\left(\frac{C}{N_0}\right)_T^{-1} = \left(\frac{C}{N_0}\right)_D^{-1} + \left(\frac{C}{N_0}\right)_U^{-1} + \left(\frac{C}{N_0}\right)_{IM}^{-1} + \left(\frac{C}{N_0}\right)_I^{-1} \qquad (3.40)$$

where $\left(\frac{C}{N_0}\right)_{IM}^{-1}$ represents the inverse of the carrier to inter-modulation noise ratio and $\left(\frac{C}{N_0}\right)_I^{-1}$ the inverse of the carrier-to-interference ratio.

3.3.2.5 NTN Link Budget Examples

In this section, we report some link budget examples based on the 3GPP NTN parameters, [8]. Tables 3.5 and 3.6 report the system-level satellite

Table 3.5 Satellite parameters for system-level assessment, Set-1.

Band	Parameter	GEO, 35 786 km	LEO, 600 km	LEO, 1200 km
Ka (20 GHz DL, 30 GHz UL)	Equivalent satellite antenna aperture	5 m	0.5 m	0.5 m
	EIRP density	40 dBW/MHz	4 dBW/MHz	10 dBW/MHz
	Maximum TX/RX gain	58.5 dBi	38.5 dBi	38.5 dBi
	G/T	28 dB/K	13 dB/K	13 dB/K
S (2 GHz DL/UL)	Equivalent satellite antenna aperture (m)	22 m	2 m	2 m
	EIRP density (dBW/MHz)	59 dBW/MHz	34 dBW/MHz	40 dBW/MHz
	Maximum TX/RX gain	51 dBi	30 dBi	30 dBi
	G/T	19 dB/K	1.1 dB/K	1.1 dB/K

Source: Adapted from [8].

Table 3.6 Satellite parameters for system-level assessment, Set-2.

Band	Parameter	GEO, 35 786 km	LEO, 600 km	LEO, 1200 km
Ka (20 GHz DL, 30 GHz UL)	Equivalent satellite antenna aperture	2 m	0.2 m	0.2 m
	EIRP density	32 dBW/MHz	−4 dBW/MHz	2 dBW/MHz
	Maximum TX/RX gain	50.5 dBi	30.5 dBi	30.5 dBi
	G/T	20 dB/K	5 dB/K	5 dB/K
S (2 GHz DL/UL)	Equivalent satellite antenna aperture (m)	12 m	1 m	1 m
	EIRP density (dBW/MHz)	53.5 dBW/MHz	28 dBW/MHz	34 dBW/MHz
	Maximum TX/RX gain	45.5 dBi	24 dBi	24 dBi
	G/T	14 dB/K	−4.9 dB/K	−4.9 dB/K

Source: Adapted from [8].

parameters for payload sets 1 and 2, respectively, and Table 3.7 provides those for the UEs as per TR 38.821.

Tables 3.8–3.11 provide the uplink and downlink link budget computations for both sets of satellite parameters and in both Ka (VSAT) and S (handheld) band. With respect to the user bandwidth, it shall be mentioned that in Ka-band users can be assumed to have 400 MHz, in full frequency

Table 3.7 UE parameters for system-level assessment.

Parameter	VSAT	Handheld
Band	Ka	S
Transmission power	2 W	200 mW
Antenna type	0.6 m diameter aperture antenna circular pol.	2 elements array linear pol.
Maximum TX gain	43.2 dBi	0 dBi/element
Maximum RX gain	39.7 dBi	0 dBi/element
Antenna temperature	150 K	290 K
Noise figure	1.2 dB	7 dB

Source: Adapted from [8].

reuse, while in S-band they have 30 MHz, in full frequency reuse, on the downlink and 360 kHz on the uplink, with any reuse scheme. In this context, it shall also be mentioned that, based on the above equations for the link budget computation, the results in terms of the CNR from (3.37) are the same for any frequency reuse scheme, since both the noise and the EIRP are scaled by the available bandwidth. While an extensive discussion on the frequency allocations is reported in Chapter 7, it is worthwhile highlighting that deployments in FR2 (Ka-band for VSAT) are easier thanks to the connection typically in line of sight, and the related absence of multi-path effects, and to the larger receiver antenna gain compared to handheld terminals, allowing to close the link budget.

For the elevation angle, as per TR 38.821, we considered 12.5° for GEO systems and 30° for LEO. Finally, the coupling loss is defined as the overall path loss to which the receiving and transmitting antenna gains are subtracted, i.e.:

$$CL = L - (G_{TX} + G_{RX}) \tag{3.41}$$

with L being the total path loss.

By comparing the performance with Set-1 and Set-2, both in the uplink and the downlink, it can be noticed that the CNR is approximately 8 and 5 dB lower with the latter; this is in line with the payload configuration reported in Tables 3.5 and 3.6. In fact, Set-2 parameters correspond to significantly less directive antennas on-board the satellite and to lower available EIRP densities. Thus, while the VSAT (Ka-band) and handheld (S-band) terminals have the same characteristics, the payload offers a worse performance with Set-2. In terms of constellation orbits, it can be observed that GEO satellites

Table 3.8 NTN DL link budget with Set-1 parameters.

Parameter	Units	Ka-band			S-band		
		GEO	LEO	LEO	GEO	LEO	LEO
Altitude	km	35 786	600	1200	35 786	600	1200
EIRP density $EIRP_d$	dBW/MHz	40	4	10	59	34	40
User bandwidth B	MHz	400			30		
EIRP: $EIRP_d + 10\log_{10} B$ (MHz)	dBW	66.02	30.02	36.02	73.77	48.77	54.77
Elevation angle ε	deg	12.5	30	30	12.5	30	30
Slant range d	km	40 316.68	1075.09	1998.88	40 316.68	1075.09	1998.88
Carrier frequency	GHz	20			2		
Free space loss L_{FS}	dB	210.58	179.10	184.49	190.58	159.10	164.49
Additional loss L_{ad}	dB	0			0		
Scintillation loss L_{scint}	dB	1.08	0.3	0.3	2.2		
Gaseous absorption L_{gas}	dB	1.2	0.52	0.52	0.16	0.07	0.07
Depolarization loss L_{pol}	dB	0			0		
Shadowing margin L_{sha}	dB	0			3		
Total path loss L $L_{FS} + L_{ad} + L_{scint} + L_{gas} + L_{pol} + L_{sha}$	dB	212.86	179.92	185.31	195.94	164.37	169.76
TX antenna gain G_{tx}	dBi	58.5	38.5		51	30	
RX antenna gain G_{rx}	dBi	39.7			0		
Noise figure N_f	dB	1.2			7		
RX antenna temperature T_a	K	150			290		
RX equivalent antenna temperature T from (33)	dBK	23.84			31.62		
Receiver $G/T = G_{rx} - T$	dB/K	15.86			−31.62		
Boltzmann constant κ	dBW/K/Hz	−228.6					
Coupling loss $L - (G_{tx} + G_{rx})$	dB	114.66	101.72	107.11	144.94	134.37	139.76
CNR $EIRP + G/T - L - 10\log_{10} \kappa B$ (Hz)	dB	11.6	8.54	9.15	0.03	6.61	7.22

Table 3.9 NTN DL link budget with Set-2 parameters.

Parameter	Units	Ka-band			S-band		
		GEO	LEO	LEO	GEO	LEO	LEO
Altitude	km	35 786	600	1200	35 786	600	1200
EIRP density $EIRP_d$	dBW/MHz	32	−4	2	53.5	28	34
User bandwidth B	MHz		400			30	
EIRP: $EIRP_d + 10\log_{10}B$ (MHz)	dBW	58.02	22.02	28.02	68.27	42.77	48.77
Elevation angle ϵ	deg	12.5	30	30	12.5	30	30
Slant range d	km	40 316.68	1075.09	1998.88	40 316.68	1075.09	1998.88
Carrier frequency	GHz		20			2	
Free space loss L_{FS}	dB	210.58	179.10	184.49	190.58	159.10	164.49
Additional loss L_{ad}	dB		0			0	
Scintillation loss L_{scint}	dB	1.08	0.3	0.3		2.2	
Gaseous absorption L_{gas}	dB	1.2	0.52	0.52	0.16	0.07	0.07
Depolarization loss L_{pol}	dB		0			0	
Shadowing margin L_{sha}	dB		0			3	
Total path loss L $L_{FS}+L_{ad}+L_{scint}+L_{gas}+L_{pol}+L_{sha}$	dB	212.86	179.92	185.31	195.94	164.37	169.76
TX antenna gain G_{tx}	dBi	50.5	30.5		45.5	24	
RX antenna gain G_{rx}	dBi		39.7			0	
Noise figure N_f	dB		1.2			7	
RX antenna temperature T_a	K		150			290	
RX equivalent antenna temperature T from (33)	dBK		23.84			31.62	
Receiver $G/T = G_{rx} - T$	dB/K		15.86			−31.62	
Boltzmann constant κ	dBW/K/Hz			−228.6			
Coupling loss $L - (G_{tx}+G_{rx})$	dB	122.66	109.72	115.11	150.44	140.37	145.76
CNR $EIRP + G/T - L - 10\log_{10}\kappa B$ (Hz)	dB	3.6	0.54	1.15	−5.47	0.61	1.22

Table 3.10 NTN UL link budget with Set-1 parameters.

Parameter	Units	Ka-band GEO	Ka-band LEO	Ka-band LEO	S-band GEO	S-band LEO	S-band LEO
Altitude	km	35 786	600	1200	35 786	600	1200
TX power P_{tx}	dBW	3	3	3	-7	-7	-7
User bandwidth B	MHz	400	400	400	0.36	0.36	0.36
TX antenna gain G_{tx}	dBi	43.2	43.2	43.2	3	3	3
EIRP: $P_{tx}+G_{tx}$	dBW	46.2	46.2	46.2	-4	-4	-4
Elevation angle ε	deg	12.5	30	30	12.5	30	30
Slant range d	km	40 316.68	1075.09	1998.88	40 316.68	1075.09	1998.88
Carrier frequency	GHz	30	30	30	2	2	2
Free space loss L_{FS}	dB	214.1	182.62	188.01	190.58	159.10	164.49
Additional loss L_{ad}	dB	0	0	0	0	0	0
Scintillation loss L_{scint}	dB	1.08	0.3	0.3	2.2	2.2	2.2
Gaseous absorption L_{gas}	dB	1.16	0.5	0.5	0.16	0.07	0.07
Depolarization loss L_{pol}	dB	0	0	0	0	0	0
Shadowing margin L_{sha}	dB	0	0	0	3	3	3
Total path loss $L\,L_{FS}+L_{ad}+L_{scint}+L_{gas}+L_{pol}+L_{sha}$	dB	216.34	183.42	188.81	195.94	164.37	169.76
RX antenna gain G_{rx}	dBi	58.5	38.5	38.5	51	30	30
Receiver $G/T = G_{rx}-T$	dB/K	28	13	13	19	1.1	1.1
Boltzmann constant κ	dBW/K/Hz	-228.6					
Coupling loss $L-(G_{tx}+G_{rx})$	dB	114.64	101.72	107.11	141.94	131.37	136.76
CNR $EIRP+G/T-L-10\log_{10}\kappa B$ (Hz)	dB	0.44	18.35	12.97	-7.90	5.77	0.38

Table 3.11 NTN UL link budget with Set-2 parameters.

Parameter	Units	Ka-band GEO	Ka-band LEO	Ka-band LEO	S-band GEO	S-band LEO	S-band LEO
Altitude	Km	35 786	600	1200	35 786	600	1200
TX power P_{tx}	Dbw	3			-7		
User bandwidth B	MHz	400			0.36		
TX antenna gain G_{tx}	DBi	43.2			3		
EIRP: $P_{tx}+G_{tx}$	Dbw	46.2			-4		
Elevation angle ϵ	Deg	12.5	30	30	12.5	30	30
Slant range d	Km	40 316.68	1075.09	1998.88	40 316.68	1075.09	1998.88
Carrier frequency	GHz	30			2		
Free space loss L_{FS}	Db	214.1	182.62	188.01	190.58	159.10	164.49
Additional loss L_{ad}	Db			0			
Scintillation loss L_{scint}	Db	1.08	0.3	0.3	2.2		
Gaseous absorption L_{gas}	Db	1.16	0.5	0.5	0.16	0.07	0.07
Depolarization loss L_{pol}	Db			0			
Shadowing margin L_{sha}	Db	0			3		
Total path loss $L\; L_{FS}+L_{ad}+L_{scint}+L_{gas}+L_{pol}+L_{sha}$	Db	216.34	183.42	188.81	195.94	164.37	169.76
RX antenna gain G_{rx}	DBi	50.5	30.5		45.5	24	
Receiver $G/T=G_{rx}-T$	dB/K	20	5	5	14	-4.9	-4.9
Boltzmann constant κ	dBW/K/Hz			-228.6			
Coupling loss $L-(G_{tx}+G_{rx})$	dB	122.64	109.72	115.11	147.44	137.37	142.76
CNR $EIRP+G/T-L-10\log_{10}\kappa B$ (Hz)	dB	-7.56	10.36	4.97	-12.90	-0.23	-5.62

allow to close the link budget in Ka-band downlink with both configuration sets, while in all other conditions, the slant range is too large to compensate for the reduced antenna directivity and emission power levels. With respect to LEO systems, the link budget poses less challenges. In general, it can be observed that: (i) on the downlink, Set-1 provides better performance, while Set-2 would allow lower data rates (corresponding to CNR values between 0 and 2 dB); and (ii) on the uplink, the link budget can be closed with a good performance in general, except for S-band systems configured with Set-2 (which is the worst scenario in terms of antenna directivity) and S-band systems with Set-1 when the LEO orbit is higher (i.e. 1200 km or above).

References

1 H. D. Curtis, *Orbital Mechanics for Engineering Students*, Elsevier Butterworth-Heinemann, 1st ed. 2005.

2 J. R. Wertz, W. J. Larson, *Space Mission Analysis and Design*, Space Technology Library, 3rd ed. 1992.

3 B. G. Evans, *Satellite Communication Systems*, IET Telecommunication Series 38, 3rd ed., 2008.

4 H. Cottin et al., "Space as a tool for astrobiology: review and recommendations for experimentations in earth orbit and beyond", *Space Science Reviews*, vol. 209, pp. 83–181, 2017.

5 G. Maral, M. Bousquet, *Satellite Communications Systems – Systems, Techniques and Technologies*, Wiley, 5th ed., 2009.

6 J. G. Walker, "Some circular orbit patterns providing continuous whole Earth coverage", *Journal of the British Interplanetary Society*, vol. 24, pp. 369–384, 1971.

7 J. G. Walker, "Satellite constellations", *Journal of the British Interplanetary Society*, vol. 37, pp. 559–571, 1984.

8 3GPP TR 38.821 V16.1.0: "Solutions for NR to support non-terrestrial networks (NTN) (Release 16)", May 2021.

9 3GPP TSG RAN1 meeting #106-e, Thales: "Considerations on UL timing and frequency synchronization in NTN", August 2021.

10 3GPP TR 38.811 V15.4.0: "Study on New Radio (NR) to support non-terrestrial networks (Release 15)", September 2020.

11 A. Vanelli-Coralli et al., "5G and Beyond 5G Non-Terrestrial Networks: trends and research challenges", *2020 IEEE 3rd 5G World Forum (5GWF)*, pp. 163–169, 2020.

12 A. Guidotti et al., "Architectures and key technical challenges for 5G systems incorporating satellites", *IEEE Transactions on Vehicular Technology*, vol. 68, no. 3, pp. 2624–2639, March 2019.

13 A. Guidotti et al., "Architectures, standardisation, and procedures for 5G Satellite Communications: a survey", *Computer Networks*, vol. 183, December 2020.

14 A. Guidotti et al., "Satellite-enabled LTE systems in LEO constellations", *2017 IEEE International Conference on Communications Workshops (ICC Workshops)*, pp. 876–881, May 2017.

15 ITU-R Recommendation P.618-13: *"Propagation data and prediction methods required for the design of Earth-space telecommunication systems"*, December 2017.

4

NR NTN Architecture and Network Protocols

4.1 Introduction

The journey of non-terrestrial networks (NTN) toward The 3rd Generation Partnership Project (3GPP) started years ago in Rel-15 with the first study on scenarios and channel models [1], followed by another study on architecture in Rel-16 [2]. These two studies paved the way for the introduction of NTN in 3GPP specifications in Rel-17. Rel-17 is then the first 3GPP release that supports NTN. In the same release, NTN support for Enhanced Universal Terrestrial RAN (E-UTRAN) Internet of Things (IoT) and Long Term Evolution (LTE) Cat-M (LTE-M) is also added following a quick study [3], taking as baseline the same NTN architecture and signaling introduced in New Radio (NR). Specifications of NR IoT are foreseen for Rel-18, to be based on the same NTN architecture and signaling.

By shifting from a vertically integrated and proprietary architecture, protocol stack and radio layer toward globally standardized ones, NTN are becoming part of the 3GPP ecosystem. This will promote interoperability in general and will pave the way for a broader global market. It is then clear that the NR NTN architecture as specified by 3GPP provides a solid basis for current and future functionality with such goals in mind.

In this chapter, we will present the NR NTN architecture, focusing on what is specified in Rel-17 but also mentioning other options considered during past studies (especially, regenerative architectures). Some such regenerative architectures show promise for future evolutions of NTN.

4.2 Architecture Overview

NTN as specified by 3GPP in Rel-17 is shown in Figure 4.1 [4]. Non-terrestrial NR access to the User Equipment (UE) is provided by means of

5G Non-Terrestrial Networks: Technologies, Standards, and System Design, First Edition.
Alessandro Vanelli-Coralli, Nicolas Chuberre, Gino Masini, Alessandro Guidotti, and Mohamed El Jaafari.
© 2024 The Institute of Electrical and Electronics Engineers, Inc. Published 2024 by John Wiley & Sons, Inc.

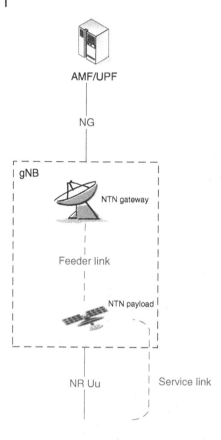

Figure 4.1 NTN in 3GPP Rel-17.
Source: Adapted from [4].

an NTN payload and an NTN Gateway: the service link is between the NTN payload and the UE, and the feeder link is between the NTN Gateway and the NTN payload. The gNB, therefore, includes the NTN payload and the NTN gateway.

The architecture that has been selected is *transparent*: the NTN payload transparently forwards the radio signal received from the UE via the service link to the NTN Gateway via the feeder link, and vice versa. In other words, the NTN payload is specified as a bent pipe. Because of this, a gNB may serve multiple NTN payloads, and an NTN payload may be served by multiple gNBs. The service link and the feeder link may operate on different frequencies, so the NTN payload is allowed to change carrier frequency before re-transmitting the signals between the two links.

Also, for NTN, the concept of *Tracking Area* (TA) applies. A TA is defined as a set of cells where a UE can move around without updating the core network. The network allocates one or more TAs to a UE. In NTN, a TA corresponds to a fixed geographical area. NTN cells can be both fixed or moving with respect to the ground (for example, in case of non-geostationary systems), in which case a suitable mapping is configured consistently in the Radio Access Network (RAN) and the 5th Generation Core Network (5GC). This mapping enables the network to pinpoint the "real" area where the UE happens to be located.

Notice that the only network interface shown in Figure 4.1 is NG. The standard *per se* does not preclude deploying Xn between NTN gNBs: from a geographical point of view, Xn would run between sites where NTN Gateways (GWs) are located. But these sites may be located hundreds or thousands of kilometers apart, and they typically do not require tight coordination for their operation, unlike terrestrial gNBs.[1] Then inter-gNB mobility or dual connectivity via Xn between NTN GWs, while not precluded by the 3GPP standard, would not be very relevant, for at least two reasons. First, because it would be unlikely that two gNBs so far apart would be connected to the same core network; second, because the advantage of Xn-based mobility with respect to NG-based mobility when source and target gNBs are so far apart, would be nonexistent.

Similar considerations apply to DC, which requires Xn to be available between the master and secondary nodes (see Figure 4.2).

DC requires tight coordination of radio operation between the two involved nodes and imposes requirements on user plane packet buffering over Xn between the two. There would be no benefit in using dual connectivity in this situation, when the two nodes are so far apart.

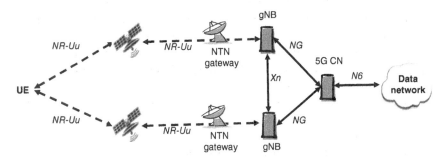

Figure 4.2 Dual connectivity with Rel-17 NTN architecture [2].

---

1 In terrestrial networks, Xn is typically deployed within local "clusters" of gNBs, which share the same core network to support, among others (see Section 4.2.2): optimized inter-gNB mobility (i.e., without the need to change core network anchor), dual connectivity, load balancing, and tight coordination of radio parameters.

Among the benefits of this architecture is that it enables early deployment of NTNs with the lowest possible complexity for the satellite payload. In principle, it is possible to deploy a Rel-17 NTN even with an existing satellite, as long as it can support the correct frequency band for the NR-Uu air interface toward the UEs.

4.2.1 NG Interface Functions

The NG interface supports the following functions [5]:

- Paging: Sending of paging requests to the New (G) RAN (NG-RAN) nodes involved in the paging area (belonging to the TA where the UE is registered)
- UE context management: Allowing the Access and Mobility Function (AMF) to establish, modify, or release a UE context in both the RAN and the core network
- Mobility management: Supporting mobility both within NG-RAN and inter-system
- Packet Data Unit (PDU) session management: Establishing, modifying, and releasing the NG-RAN resources for the involved PDU sessions once an UE context is established
- Non-Access Stratum (NAS) transport: To transport, reroute, or report lack of delivery of NAS messages (transparent signaling between the UE and the AMF)
- NAS node selection: To select the appropriate AMF for the UE, also taking into account the UE location
- Interface management: To set up the interface and update configuration information between the RAN and the core network
- Warning message transmission: Supporting the broadcast of warning messages or canceling an ongoing broadcast for Public Warning System (PWS) operation
- Configuration transfer: Transferring RAN configuration information (e.g., Self-Organizing Networks [SON] information) between RAN nodes through the core network
- Trace: To control UE trace sessions in the RAN
- AMF management: Supporting AMF planned removal and auto-recovery
- Multiple Transport Network Layer (TNL) associations support: Selecting and releasing a TNL association between an AMF and the NG-RAN node (based on usage and weight factor)
- AMF load balancing: The AMF can indicate its relative capacity to the NG-RAN for load-balancing purposes

- Location reporting: The AMF can request the current location, or the last known location with timestamp, or the presence in a certain area of interest, for a UE
- AMF re-allocation: Redirecting an initial connection request from an initial AMF to a different one
- UE radio capability management: To handle UE radio capability information
- NR Positioning Protocol a (NRPPa) signaling transport: Transparent transport of NRPPa messages between the NG-RAN and the LMF through the AMF to support UE positioning functionality
- Overload control: The AMF can control the load generated by the NG-RAN nodes connected to it
- Secondary Radio Access Technology (RAT) data volume reporting: To report secondary RAT data volume usage for dual connectivity
- Remote Interference Management (RIM) information transfer: Transferring RIM information between two RAN nodes through the core network
- Radio Access Network Control Plane (RAN CP) relocation: Relocating a UE-associated connection for an NarrowBand IoT (NB-IoT) UE using CP CIoT 5GS optimization, allowing the re-establishment to be re-authenticated by the AMF
- Suspend-Resume: To suspend a UE-associated connection and release the UP tunnel while storing the context in RAN for a subsequent resume (only for long Enhanced Discontinuous Reception [eDRX] cycles)

4.2.2 Xn Interface Functions

We have previously observed that Xn deployment with NTN in this architecture does not seem practical, but it is not precluded by the standard. For the sake of completeness, it seems beneficial to also list the functions supported by Xn [6]. For control plane:

- Interface management and error handling: To set up, reset, remove Xn, indicate errors, and update configuration data between two NG-RAN nodes
- UE mobility management: To prepare and cancel UE handovers, retrieve UE context between NG-RAN nodes, page a UE in inactive state, and control data forwarding
- Dual connectivity: To set up, modify, and release radio resources for concurrent operation toward the same UE between a master and a secondary node
- Energy saving: Indicating cell activation/deactivation between NG-RAN nodes

- Resource coordination: To coordinate cell resource usage
- Secondary RAT data volume reporting: To report secondary RAT usage from the secondary to the master node for DC operation
- Trace: To control trace sessions for a UE
- Load management: Exchanging resource status and load information between NG-RAN nodes
- SON data exchange: Exchanging information for SON support

And for user plane:

- Data transfer: Transferring User Plane (UP) data between NG-RAN nodes for mobility or dual connectivity
- Flow control: The NG-RAN node that receives UP data from another NG-RAN node can feed back information associated with the data flow
- Assistance information: The NG-RAN node that receives UP data from another NG-RAN node can feed back assistance information (e.g. related to radio conditions) to the other node
- Fast retransmission: For dual connectivity, in case of radio outage in one of the nodes, to handle in the node in good radio conditions any data previously forwarded to the node in outage

4.3   User Plane and Control Plane

Figure 4.3 shows the structure of an NTN, showing the relationship among Radio Bearers, NG-U Tunnels, PDU sessions, and Quality of Service (QoS) Flows, with the end-to-end mapping of QoS flows in evidence. This follows the same principles and rules for traffic handling in the NG-RAN and the 5GC. The satellite and the NTN gateway (the blue boxes in Figure 4.3) make up the gNB in the logical architecture view. PDU sessions and QoS flows are terminated in the UE and the UPF.

Control plane and user plane stacks for NTN do not change with respect to the terrestrial case.

4.3.1   Control Plane

The control plane stack is shown in Figure 4.4. Both the satellite payload and the NTN gateway are transparent with respect to the CP protocols, so the termination points for Radio Resource Control (RRC), Packet Data Convergence Protocol (PDCP), Radio Link Control (RLC), Medium Access Control (MAC), and PHYsical layer (PHY) are on the UE and the gNB. NAS PDUs are terminated by the AMF in the core network and by the UE.

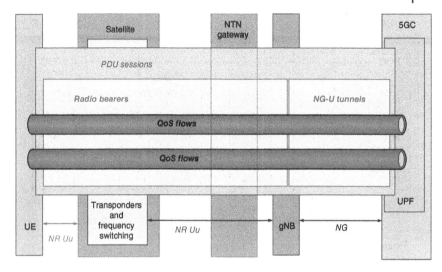

Figure 4.3 NTN based on a transparent satellite, showing end-to-end mapping of QoS flows [2].

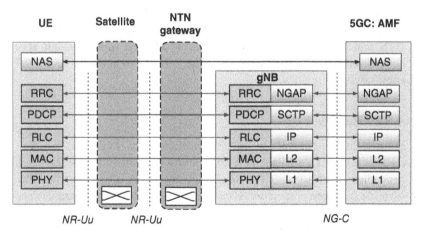

Figure 4.4 Control plane stack for NTN [2].

4.3.2 User Plane

The user plane stack is shown in Figure 4.5. Both the satellite payload and the NTN gateway are transparent with respect to the UP protocols, so the termination points for Service Data Adaptation Protocol (SDAP), PDCP, RLC, MAC, and PHY are on the UE and the gNB. UP PDUs are terminated by the UPF in the core network and by the UE.

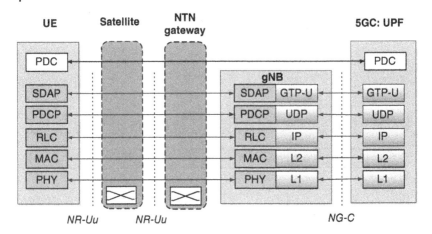

Figure 4.5 User plane stack for NTN [2].

4.4 Interworking with Terrestrial Mobile Networks

We have seen the functions supported by NG-RAN network interfaces in Rel-17 in Sections 4.2.1 and 4.2.2. In principle, interworking between NTN and terrestrial networks can involve all such functions, but special considerations must be made with respect to mobility and dual connectivity.

4.4.1 Mobility

UE mobility to and from terrestrial networks leverages the NG interface (NG mobility). This means that all mobility procedures follow the same principles as terrestrial 3GPP Rel-17 NR, regardless of whether mobility involves the same core network nodes (intra-AMF case) or different ones (inter-AMF case).

As discussed in Section 4.2, the use of Xn is not precluded from the point of view of the standards; however, from a practical point of view deploying Xn between NTN and terrestrial gNBs would imply the following:

- Connecting NTN and terrestrial gNBs to the same core network;
- Close interworking between NTN and the involved gNBs.

While the first condition can be met, the second would require the NTN gNB and the involved gNBs to be considered "neighbors." Given that an NTN gNB in this scenario can potentially serve a whole continent, it is not

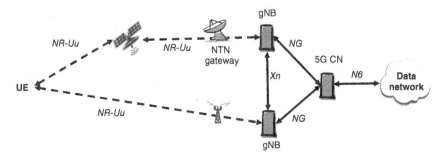

Figure 4.6 Dual connectivity involving NTN and terrestrial networks [2].

realistic to consider such a large set of terrestrial gNbs, covering such a large area, as "neighbors" with which to coordinate, e.g., mobility or radio parameters. From a geographical point of view, we can make the same observations as in Section 4.2: deploying Xn over such a large geographical area outsteps the typical use case for such interface. Even restricting to a smaller subset of potential terrestrial gNbs (e.g., those in the vicinity of the NTN GW) would not be a meaningful scenario. In conclusion, it does not seem practical that an operator (NTN or otherwise) would deploy Xn interfaces from an NTN GW site to terrestrial gNbs with this architecture.

4.4.2 Dual Connectivity

As previously discussed, dual connectivity in general requires a horizontal network interface (Xn or X2) to be deployed; furthermore, dual connectivity puts additional constraints on user plane buffering. When considering possible dual connectivity between NTN and terrestrial networks with Rel-17 (see Figure 4.6), the same observations can be made for Xn mobility (Section 4.4.1).

Although not precluded by the standard, dual connectivity between terrestrial and NTN does not seem practical with this architecture.

4.5 Impact on Other Technologies: IoT NTN

The 3GPP Rel-17 NTN architecture has also been adopted for Rel-17 IoT/Machine-type Communications (MTC) NTN, reusing it in the scope of E-UTRAN and limited to this functionality [7]. This is shown in Figure 4.7 below. The same considerations made in earlier sections apply, with some notable limitations due to the specific characteristics of IoT and MTC [8].

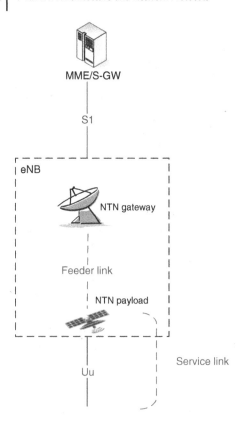

Figure 4.7 NTN for E-UTRAN
IoT/MTC in 3GPP Rel-17.
Source: Adapted from [7].

4.6 Regenerative Architectures

One of the reasons behind the adoption of the architecture with a transparent payload described above, is that it enables early implementations and early deployment of NTNs. It limits the complexity of the satellite payload, which is essentially a bent pipe: the satellite payload repeats the air interface (Uu) from the feeder link to the cells on the ground, and vice versa. At least in principle, even an existing payload already in operation could be reused for Rel-17 NTN if it supports the correct frequency bands with the appropriate RF parameters. But if we look into the future beyond Rel-18, an architecture based on a regenerative satellite payload seems very beneficial for 5G (and even potential 6G) services. Among the benefits of regenerative payloads is the possibility to implement packet switching at various levels directly in the payload, without the need to go through the ground segment, leveraging inter-satellite links if present; such capability is added on top of what is provided by the ground segment itself, for additional capacity and resiliency. Regenerative payloads also unlock the possibility to move at least some core network functionality on board, for even more added resiliency in case connectivity to the ground station is lost.

Some such architectures have been studied by 3GPP and are documented in [2]; they involve moving all or part of the gNB functionality on board the satellite. We will describe some of them, starting with the one that seems to be the most promising.

4.6.1 NG-RAN Node on Satellite

Putting a full NG-RAN node on a satellite seems like an obvious evolution of the NTN architecture. Terminations for all the network interfaces then reside on the satellite, and the NTN gateway on the ground only remains as a transport node. The feeder link then transports the NG interface toward the core network, and Inter-Satellite Links (ISLs) carry the Xn interface between satellite-carried gNBs. This is shown in Figure 4.8 [2]. gNBs on board different satellites may be connected to the same core network on the ground. As the gNB is a logical node, it is not precluded that the satellite hosts more than one gNB, in which case the same Satellite Radio Interface (SRI) transports all the corresponding NG interface instances.

The presence of Xn between satellites unlocks the possibility of supporting more advanced features within the NTN, leveraging the currently specified Xn functions (see the list in Section 4.2.2). Some examples that seem particularly advantageous are listed below.

● Inter-satellite mobility: UEs can be handed over from one satellite to another without involving the core network, unlike with the Rel-17/18 architecture.

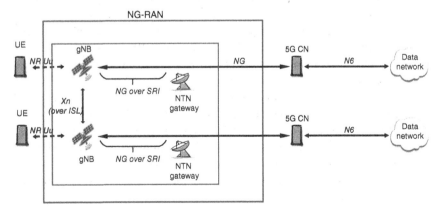

Figure 4.8 Regenerative satellite with ISL [2].

- Inter-satellite dual connectivity: NTN gNBs may act as MNs or SNs, leveraging the Xn, to provide additional radio resources toward the UE.
- Resource coordination, SON: NTN gNBs are able to coordinate radio resources in order to optimize their capacity and performance.
- Energy saving: NTN gNBs are able to signal cell (beam) activation/ deactivation to one another.

Similarly to the Rel-17/18 architecture with a transparent payload, also in this case running Xn between a satellite and a terrestrial gNB is not precluded. However, the same considerations as in Section 4.4.1 apply for this architecture: Xn between two NG-RAN nodes is deployed if they are connected to the same core network, allowing for close interworking between the involved nodes. Also in this case, it seems questionable to consider a satellite-based gNB and an arbitrary terrestrial gNB in its coverage area as "neighbors," which would benefit from coordinating their parameters in a peer-to-peer manner through a horizontal interface.

The fact that this architecture requires to host the gNB on board the satellite is also one of its drawbacks: it is, in fact, the most complex in terms of satellite payload implementation. But given that it enables support in NTN the full range of currently specified 3GPP functionality, it is arguably also the most future-proof.

4.6.2 Split Architectures

The impact on payload implementation of the architecture described in Section 4.6.1 is undoubtedly high, as the payload is required to host a full gNB, including the termination of all its network interfaces. By considering hosting only a subset of gNB functionality in the satellite payload while the rest stays on the ground at the NTN GW site, it might be possible to reduce its complexity. At least two possibilities could be considered: either hosting only the gNB-DU on board while leaving the gNB-Central Unit (CU)-CP and gNB-CU-UP on the ground (following the NG-RAN split architecture specified by 3GPP) or hosting only the lower layer part of the gNB-DU on board (following the lower-layer split adopted by the O-RAN alliance [9]). As we will see both alternatives have their pros and cons, but in any case, splitting the gNB in an NTN scenario comes at a cost, considering that these split architectures have not been specified with the NTN scenario in mind, and may impose additional constraints to the NTN use case. The question is whether this is an acceptable price to pay in exchange for a simpler payload implementation.

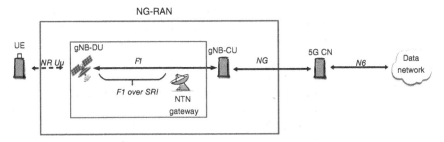

Figure 4.9 NG-RAN with a regenerative satellite based on gNB-DU [2].

4.6.2.1 CU–DU Split

In Chapter 2, we have seen that NG-RAN supports splitting the gNB into its components: the gNB-CU-CP, the gNB-CU-UP, and the gNB-DU. This can be used as basis also for NTN, with the gNB-DU on board the satellite and the gNB-CU-CP and gNB-CU-UP (which make up the gNB-CU) on the ground. This is shown in Figure 4.9 [2].

The satellite payload implements regeneration of the signals received from Earth. The NR Uu radio interface is on the service link between the satellite and the UE; the satellite radio interface on the feeder link between the NTN GW (which hosts the gNB-CU) and the satellite transports the F1 interface specified by 3GPP. Consistently with the cardinality specified in the NG-RAN split architecture, gNB-Distributed Unit (DUs) on board different satellites may connect to the same gNB-Central Unit (CU) on the ground. If the satellite hosts more than one gNB-DU, the same SRI transports all the corresponding F1 interfaces.

The satellite payload may provide ISLs between satellites; however, it is not possible to support standardized horizontal interfaces between satellites: due to the functional split and hierarchy on which the CU–DU split is based, no inter-gNB–DU interface is supported in the 3GPP standard. One immediate consequence is that, with the CU–DU split architecture, it is not possible to support direct inter-satellite mobility within the standardized functional split (the only possible horizontal interface, Xn, is terminated between gNB and CUs on the ground and does not reach any satellite).

Another constraint that this architecture imposes on the NTN scenario is a consequence of the functional split adopted in the CU–DU split architecture. The F1 interface is designed to be "persistent" between the same CU/DU pair and does not support being torn down and set up again, let alone toward a different gNB-CU from the same gNB-DU. Tearing down F1 means that any UE contexts handled over that interface are released and the corresponding UE connections are going to be dropped, as F1 has no native function to relocate UE contexts "on demand" toward a different CU/DU pair. Practically speaking, this means that this architecture cannot support a LEO satellite changing NTN GWs as it moves along its orbit unless something is done in the implementation to address this limitation.

One such "clever" implementation could host two gNB-DUs on board the same satellite, simultaneously connected to different gNB-CUs on the ground. The two gNB-DUs serve simultaneously the same UEs, coordinating and sharing the relevant information with each other via an implementation-dependent internal interface. When the satellite is in the process of changing its connection from the "old" NTN GW to the "new" one all UE contexts are transferred from the gNB-DU connected to the "old" gNB-CU to the gNB-DU connected to the "new" gNB-CU. The above example is just one possible way how a payload implementation might solve this problem outside of the 3GPP specifications.

Overall, then, the price to pay to avoid a full gNB on board is additional complexity of implementation in order to overcome additional constraints that this particular architecture imposes on the NTN scenario.

4.6.2.2 Lower-layer Split

Another possibility for a regenerative satellite that does not require the full gNB on board, is to leverage the so-called Lower-layer Split (LLS), where the split between the "centralized" and "distributed" parts of the gNB happens further down in the function stack. A number of alternatives are possible in principle, and they were studied by 3GPP in preparation for the standardization of 5G [10]. The alternatives considered by 3GPP are shown in Figure 4.10 (the numbering has no relationship to the numbering of NG-RAN deployment options).

Option 2 in Figure 4.10, which calls for RRC and PDCP in the central unit, and RLC, MAC, PHY, and RF in the distributed unit, was the one selected for standardization, which resulted in the current gNB-CU, gNB-DU, and F1 functionality. Options 6 and 7 (MAC-PHY split and intra-PHY split, respectively), which call for upper layers in the central unit and RF and at least parts of PHY in the distributed unit, were the subject of a dedicated study in 3GPP [11]. One of the inner *Matryoshkas* of the gNB was then opened up to further look at its components, as shown in Figure 4.11. Three subvariants for Option 7 (see Figure 4.11) were also considered.

The 3GPP study encountered a number of challenges. For example, 3GPP specifications do not describe base station functionality (always a "moving target" due to technology improvements and different implementations); the feasibility of the different LLS options heavily depends on the

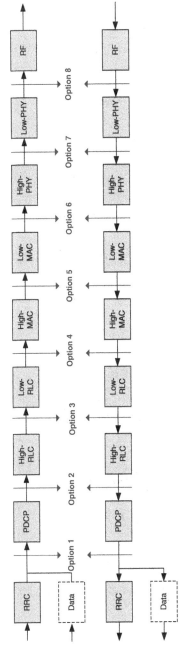

Figure 4.10 Possible split between a "central" and a "distributed" unit within the gNB [10].

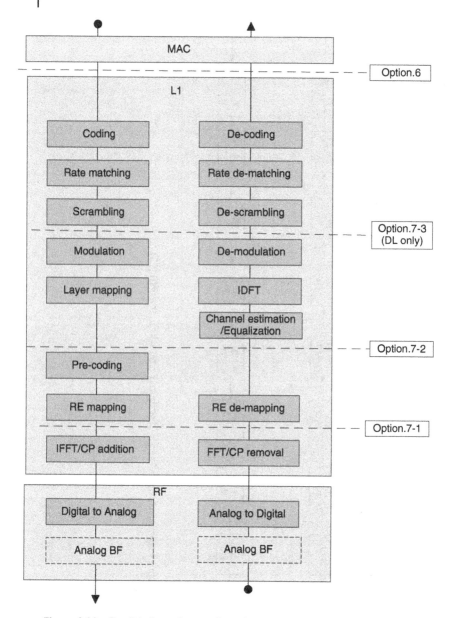

Figure 4.11 Possible lower layer split options studied in 3GPP. DL is on the left, and UL is on the right. Source: [11] (2020)/3GPP.

performance of the link used to transport the interface between central and distributed units; the required transport capacity for such a link, in turn, heavily depends on the selected radio configuration (system bandwidth, Multiple Input Multiple Output [MIMO] configuration, if any, number of antenna ports, and so on); furthermore, transport network characteristics are out of 3GPP scope. The conclusion in 3GPP was that all options considered were technically feasible [12].

The work on LLS has continued in the O-RAN Alliance, which defined an architecture where the gNB-DU is further split into DU and RU. In this architecture, the DU hosts RLC, MAC, and high PHY, and the Remote Unit (RU) hosts the low PHY [9].

Work on fronthaul specifications to support the DU–RU interface has also progressed in other Standards-Developing Organization (SDOs). Common Public Radio Interface (CPRI) released their Enhanced CPRI (ecPRI) specifications supporting NR and specified the additional LLS Option 8 (PHY-RF split), which calls for RF functionality in the distributed unit and all upper layers in the central unit. The goal behind Option 8 was to simplify the interface between the two nodes: it would have to convey mostly I/Q data samples, some control, and synchronization, although this option has the most stringent requirement on the fronthaul interface in terms of latency and especially throughput [9]. IEEE also released the 1914.3 specifications, defining Ethernet-based transport that could be used with Options 7-1 and 8.

As 3GPP initially identified during its study on LLS, the fronthaul bandwidth requirements are one of the critical factors to be considered when evaluating gNB LLS options, and they highly depend on radio configuration. In an NTN scenario, the RU would be hosted on the satellite, and the DU–RU interface would be transported by the SRI, and this makes such requirements even more critical and limiting.

A comparison of the DownLink (DL) bandwidth required for the DU–RU interface in various LLS options is shown in Table 4.1. The numbers refer to a terrestrial network configuration with 128 TX/RX massive MIMO system having 100 MHz system bandwidth, $16 + 16$ I/Q samples, 256 Quadrature Amplitude Modulation (QAM), and 30 kHz sub-carrier spacing in sub-6 GHz spectrum [9].

Table 4.1 is consistent with, and confirms, 3GPP's initial findings: even when considering a fixed reference configuration (which may or may not work in an NTN scenario), the required bandwidth for fronthaul between the DU and the RU (which translates to the SRI for a single satellite for NTN), varies by 2 orders of magnitude according to the split option chosen by the specific implementation.

Table 4.1 Bandwidth required for fronthaul transport (DL only), for various LLS options.

Split option	Bandwidth (Gbit/s)	Description
6	5.8	MAC and upper layers are in the central unit; PHY and RF are in the distributed unit
7-1	~376	iFFT, cyclic prefix addition/removal, and PRACH filtering are in the RU; all other PHY functions are in the distributed unit
7-2	~23	iFFT, cyclic prefix addition/removal, PRACH filtering, and precoding are in the RU; all other PHY functions are in the distributed unit
7-3	~5.8	(DL only) The encoder resides in the distributed unit; all other PHY functions are in the RU
8	~503	RF chains are in the RU; all other PHY functions are in the distributed unit

Source: Sirotkin [9]/John Wiley & Sons.

In summary, implementation choices pertaining to an NTN regenerative satellite architecture based on LLS are highly dependent on the attainable performance on the SRI, which in turn will translate to varying levels of complexity for the payload itself. On top of this, this option is subject to the same constraints and limitations (e.g., the lack of standardized horizontal interfaces below the gNB-CU, lack of a standardized solution for inter-satellite handover, and others), which affect other regenerative options based on the gNB split.

4.7 Conclusions

The architecture specified by 3GPP for NTN in Rel-17 and Rel-18 involves a transparent satellite payload, with the gNB deployed on the ground. We have introduced the characteristics of this architecture, including its advantages and possible disadvantages. To unlock the full potential for enhanced NTN functionality, however, it seems promising to consider adopting an architecture with a regenerative satellite payload for future system releases. We have introduced some such architectures, some of which were already studied by 3GPP in recent years, discussing their relative advantages and disadvantages.

References

1 3GPP TR 38.811: "Study on New Radio (NR) to support non-terrestrial networks", v. 15.4.0.

2 3GPP TR 38.821: "Solutions for NR to support Non-Terrestrial Networks (NTN)", v. 16.1.0.

3 3GPP TR 36.763: "Study on Narrow-Band Internet of Things (NB-IoT)/enhanced Machine Type Communications (eMTC) support for Non-Terrestrial Networks (NTN)", v. 17.0.0.

4 3GPP TS 38.300: "NR; Overall description; Stage-2", v. 17.0.0.

5 3GPP TS 38.410: "NG-RAN; NG general aspects and principles", v. 17.0.0.

6 3GPP TS 38.420: "NG-RAN; Xn general aspects and principles", v. 17.0.0.

7 3GPP TS 36.300: "E-UTRAN; Overall description; Stage-2", v. 17.0.0.

8 O. Liberg, M. Sundberg, E. Wang, J. Bergman, J. Sachs, G. Wikström, *Cellular Internet of Things – From Massive Deployments to Critical 5G Applications*, 2nd ed., Academic Press, 2019.

9 S. Sirotkin (editor), *5G Radio Access Network Architecture – The Dark Side of 5G*, Wiley, 2021.

10 3GPP TR 38.801: "Study on new radio access technology: Radio access architecture and interfaces", v. 14.0.0.

11 3GPP TR 38.816: "Study on Central Unit (CU) – Distributed Unit (DU) lower layer split for NR", v. 15.0.0.

12 X. Lin, N. Lee (editors), *5G and Beyond – Fundamentals and Standards*, Springer, 2021.

5

NR NTN Radio Interface

5.1 Introduction

The first study item on New Radio (NR) to support Non-Terrestrial Networks (NTN) was conducted in 3GPP Release-15. The objective of this study on NTN in the RAN was to establish the definition of the NTN deployment scenarios and related system parameters such as architecture, altitude, orbit, and so on, and to adapt the 3GPP channel models for NTN. The study also identified the key impact areas on the NR interface that may need further evaluation. The TSG RAN and RAN WG1 findings related to the study are collected in the [1].

Further, based on the outcomes of the Release-15 SI phase, RAN WGs have conducted in Release-16 a study on solutions for NR to support NTN. The main objective was to study a set of necessary features and adaptations enabling the operation of NR protocol in NTN with a priority on satellite access. The RAN WGs findings related to Release-16 study are collected in [2].

The specific characteristics of NTN-based system create new technical challenges compared to the ones of cellular terrestrial networks (TNs). Enhancements and adaptations enabling the operation of the NR Physical layer in NTN-based radio link should take into account the satellite's long propagation delays, large Doppler effects due to fast satellite movements, and the possible large differential delay within the same cell.

The normative work on NTN and satellite in 3GPP Technical Specification Group (TSG) Radio Access Networks (TSG RAN) was conducted as part of 3GPP Release-17. A great care has been taken to minimize impacts at UE and NG-RAN for the support of NTN through maximum reuse of existing 5G specifications [3]. This first 3GPP-defined NTN standard covers 3GPP-defined satellite access networks respectively based on the 5G NR protocols and the 4G Narrowband Internet of things (NB-IoT) and

5G Non-Terrestrial Networks: Technologies, Standards, and System Design, First Edition.
Alessandro Vanelli-Coralli, Nicolas Chuberre, Gino Masini, Alessandro Guidotti, and Mohamed El Jaafari.

enhanced Machine Type Communication (eMTC) radio protocols both operating in FR1 bands.[1]

In 3GPP Release-17 specifications, the key physical layer aspects addressed in this regard are:

- Enhancements on uplink frequency and time synchronization,
- Enhancements on features involving downlink–uplink timing relationships,
- Enhancements on Hybrid Automatic Repeat Request (HARQ),
- Other aspects such as the indication of polarization type used on the satellite side for downlink and uplink transmissions.

The focus of this chapter is the 5G NR NTN Radio Interface. Detailed descriptions of the aforementioned 3GPP Release-17 enhancements for NTN support are provided. Because access network based on Unmanned Aerial System (UAS), including High Altitude Platform Station (HAPS) could be considered a special case of non-terrestrial access with lower delay and Doppler values and variation rate, the focus in this chapter is on satellite-based NTN only.

To put the reader within the context, a brief overview of NR basic transmission schemes and a recall of main NR procedures are provided when relevant. A more detailed description of 5G radio protocols and operation, including description of 5G network architecture as well as the overall functionality of the 5G higher layers can be found in 3GPP TS 38.300 [4]. Specifications on Physical channels and modulation can be found in 3GPP TS 38.211 [5]. Specifications of Physical layer procedures can be found in 3GPP TS 38.213 [6] and 3GPP TS 38.314 for control and data, respectively.

5.2 NR Basic Transmission Scheme

5.2.1 NR Waveform

In 5G NR, the downlink transmission waveform is Orthogonal Frequency-Division Multiplexing (OFDM) using a cyclic prefix (CP-OFDM). The uplink transmission waveform is OFDM using a CP with a transform precoding function performing DFT spreading (DFT-S-OFDM), which can be either disabled or enabled. That is, CP-OFDM is applied for both downlink and uplink, while DFT-spread OFDM can also be configured for uplink.

Multiple sub-carrier spacing (SCS) (i.e., numerology) are supported in 5G NR, including 15, 30, and 60 kHz below 7.125 GHz carrier

---

1 As per RAN plenary#93 decision NR-NTN deployment in above 10 GHz bands (including FR2) and support for VSAT/ESIM NTN UE is part of NTN evolution in 3GPP Release-18.

frequencies, i.e., frequency ranges 1 or FR1 (410–7125 MHz), 60, 120, 240 kHz in FR2-1 (24 250–52 600 MHz) and 120, 480, and 960 kHz in FR2-2 (52 600–71 000 MHz) [[7], 3GPP TS 38.104]. A Numerology corresponds to one SCS in the frequency domain. It is based on exponentially scalable SCS $\Delta f = 2^\mu \times 15$ kHz, where μ defines the numerology and takes values equal to 0, 1, 3, 4, 5, and 6 for PSS, SSS, and PBCH and equal to 0, 1, 2, 3, 5, and 6 for other channels. Normal CP is supported for all SCS and extended CP is supported for $\mu = 2$.

The OFDM-based NR waveform is also used for NR NTN in Release-17. The Release-17 specifications support NR-based satellite access deployed in FR1 bands (MSS S-Band and L-Band): Only FR1 is supported with 15, 30 kHz SCS [8]. Support of higher frequency bands, i.e., above 10 GHz is part of NTN enhancements in 3GPP Release-18 [9].

OFDM is a spectral efficient waveform with several advantages and some drawbacks. A drawback of OFDM is that its multicarrier property is associated with a relatively high peak to average power ratio (PAPR) generated by summing large number of subcarriers. The PAPR depends on the waveform and the number of component carriers (CCs). For a single CC transmission in the downlink, the PAPR is 8.4 dB (99.9%). High PAPR means that high-power amplifiers (HPA) have to operate with increased back-off leading to reduced efficiency. This is not appropriate for the power-limited satellite downlink, which typically requires operation in the nonlinear region of HPA. To alleviate PAPR problems in Satellite-based NTN, PAPR-reduction algorithms that are transmitter implementation-specific can be considered [10]. Therefore, PAPR optimizations for downlink channels are not necessary to be specified for NTN as concluded in the Release-16 study on NTN (refer to [2]).

DFT-S-OFDM used in the uplink provides lower PAPR as shown in Table 5.1,[2] which is good for UEs: Transmission chain can operate with higher efficiency; closer to the power amplifier limit, and hence transmit with higher power.

Table 5.1 PAPR for DFT-spread OFDM [10].

Modulation	$\pi/2$ BPSK	QPSK	16QAM	64QAM	256QAM
PAPR (99.9%)	4.5 dB	5.8 dB	6.5 dB	6.6 dB	6.7 dB
CM (99.9%)	0.3 dB	1.2 dB	2.1 dB	2.3 dB	2.4 dB

Source: Adapted from [11].

---

2 These values are derived without spectrum shaping. With spectrum shaping, lower PAPR and cubic metric (CM) values can be obtained, e.g., 1.75 dB PAPR for $\pi/2$ BPSK.

Another drawback of OFDM waveform is its sensitivity to frequency offsets caused by Doppler shift or by frequency inaccuracies within local oscillators at the UE and gNB. Solutions for frequency synchronization in NTN with frequency pre-compensation at the UE were specified in Release-17 as will be described in Sections 5.3 and 5.4.

5.2.2 Modulation and Coding Scheme

In Downlink: The following modulation schemes are used in 5G NR for both data and higher-layer control information: QPSK, 16QAM, 64QAM, 256QAM, and 1024QAM (see TS 38.211 [5] sub-clause 7.3.1.2). For L1/L2 control, the QPSK modulation is used as specified in TS 38.211 [5] sub-clause 7.3.2.4.

In Uplink: The following modulation schemes are used in NR for both data and higher-layer control information: $\pi/2$-BPSK; when precoding is enabled, QPSK, 16QAM, 64QAM, and 256QAM (see TS 38.211 [5]) sub-clause 6.3.1.2). The modulation schemes used for L1/L2 control are BPSK, $\pi/2$-BPSK, and QPSK (see TS 38.211 [5] sub-clause 6.3.2).

5.2.3 Channel Coding

In 5G NR, for both Downlink and Uplink, the following channel coding schemes are supported:

For data: Rate 1/3 or 1/5 Low density parity check (LDPC) coding, combined with rate matching based on puncturing/repetition to achieve a desired overall code rate (For more details, see 3GPP TS 38.212 [11] sub-clauses 5.3.2). LDPC channel coder facilitates low-latency and high-throughput decoder implementations.

For L1/L2 control: For Downlink Control Information (DCI) and Uplink Control Information (UCI) size larger than 11 bits, Polar coding, combined with rate matching based on puncturing and repetition to achieve a desired overall code rate; for more details, see 3GPP TS 38.212 [11] sub-clauses 5.3.1. Otherwise, repetition for 1-bit; simplex coding for 2-bit; Reed-Muller coding for 3~11-bit DCI/UCI size.

5.2.4 NR Multiple Access Scheme

The multiple access scheme for the NR physical layer is a combination of Orthogonal Frequency-Division Multiple Access (OFDMA), Time-Division Multiple Access (TDMA), and space Division Multiple Access (SDMA). OFDMA is based on Orthogonal Frequency Division Multiplexing (OFDM)

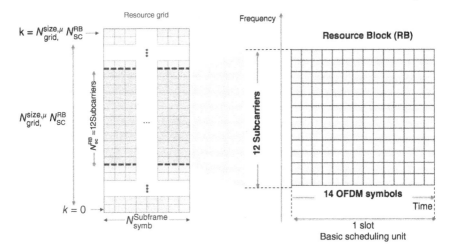

Figure 5.1 NR Resource Block (RB).

with a CP; the transmission to/from different UEs uses mutually orthogonal frequency assignments. With granularity in frequency domain assignment equal to one resource block (RB) consisting of 12 subcarriers as illustrated by the grid given in Figure 5.1.

For each numerology and carrier, a 5G NR resource grid is defined as shown in the chart in Figure 5.1. The physical dimension (i.e. SCS, number of OFDM symbols within a radio frame) varies in NR depending on numerology.

5G NR supports both duplexing methods: Frequency Division Duplex (FDD), and Time Division Duplexing (TDD). The duplexing mode in NR-based satellite systems is FDD, which allows operation on a paired spectrum. TDD operation on an unpaired spectrum can be supported in case of HAPS.

5.2.5   NR Frame Structure

The 5G NR radio frame has a fixed duration of 10 ms. It consists of 10×1 ms subframes as illustrated in Figure 5.2. Each radio frame is divided into two equally sized half-frames of five subframes each. with half-frame 0 consisting of subframes 0–4 and half-frame 1 consisting of subframes 5–9. Each subframe consists of an OFDM SCS-dependent number of slots. Each slot consists of 14 OFDM symbols (12 OFDM symbols in case of extended CP).

NR radio frames are indexed using the System Frame Number (SFN). SFN ranges from 0 to 1023: SFN cycles every 10.24 seconds.

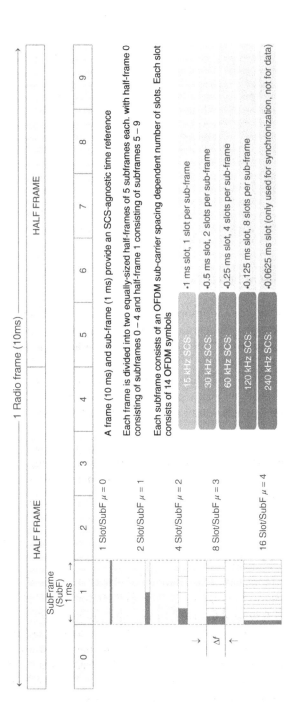

Figure 5.2 NR Frame structure.

A frame (10 ms) and sub-frame (1 ms) provide an SCS-agnostic time reference

Each frame is divided into two equally-sized half-frames of 5 subframes each. with half-frame 0 consisting of subframes 0 – 4 and half-frame 1 consisting of subframes 5 – 9

Each subframe consists of an OFDM sub-carrier spacing dependent number of slots. Each slot consists of 14 OFDM symbols

15 kHz SCS: •1 ms slot, 1 slot per sub-frame

30 kHz SCS: •0.5 ms slot, 2 slots per sub-frame

60 kHz SCS: •0.25 ms slot, 4 slots per sub-frame

120 kHz SCS: •0.125 ms slot, 8 slots per sub-frame

240 kHz SCS: •0.0625 ms slot (only used for synchronization, not for data)

1 Slot/SubF $\mu = 0$

2 Slot/SubF $\mu = 1$

4 Slot/SubF $\mu = 2$

8 Slot/SubF $\mu = 3$

16 Slot/SubF $\mu = 4$

5.2.6 Bandwidth Part Operation

In NR, a subset of the total Carrier bandwidth of a cell is referred to as a Bandwidth Part (BWP); it is a subset of contiguous common RBs for a given numerology. The UE is configured with a carrier BWP that defines the UE's operating bandwidth within the cell's operating bandwidth. For initial access, and until the UE's configuration in a cell is received, initial BWP detected from system information (SI) is used. The UE may be configured with several carrier BWPs, of which only one can be active on a given CC.

5.2.7 NR Radio Channels

NR physical layer offers service to the MAC sublayer transport channels. Physical channels are used to transfer data (MAC PDU) across radio interfaces. Transport channels are used to transfer MAC PDU (a.k.a. Transport blocks [TBs]) between the MAC and Physical layer. The MAC sublayer offers service to the RLC sublayer logical channels. Logical Channels are used to transfer RLC PDU between the RLC and MAC layers. The RLC sublayer offers service to the PDCP sublayer RLC channels. The PDCP sublayer offers service to the SDAP and RRC sublayer radio bearers: data radio bearers (DRB) for user plane data and signaling radio bearers (SRB) for control plane data. The SDAP sublayer offers 5GC QoS flows and DRBs mapping functions.

In the Downlink and Uplink, logical channels are classified into two groups: Control Channels and Traffic Channels:

- Control channels:
  - Broadcast Control Channel (BCCH): A downlink channel for broadcasting system control information: Both the Master Information Block (MIB) and System Information Blocks (SIB). The MIB is mapped onto the BCH and PBCH, whereas the SIB is mapped onto the DL-SCH and PDSCH.
  - Paging Control Channel (PCCH): a downlink channel that transfers paging messages and SI change notifications. All paging messages are mapped onto the PCH and PDSCH.
  - Common Control Channel (CCCH): Channel for transmitting control information between UEs and networks. And Dedicated Control Channel (DCCH): a point-to-point bi-directional channel that transmits dedicated control information between a UE and the network. This control information corresponds to RRC signaling messages, i.e., data belonging to the set of SRB. For the Downlink, all SRB data is mapped onto the DL-SCH and PDSCH. For the Uplink, all SRB data is mapped onto the UL-SCH and PUSCH.

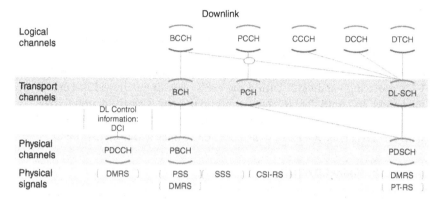

Figure 5.3 NR Downlink channels.

- Traffic channels: Dedicated Traffic Channel (DTCH) is used to transfer application data. DTCH can exist in both UL and DL.

The mappings between DL logical Channels, DL transport channels, and DL PHY channels are shown in Figure 5.3.

A downlink physical channel corresponds to a set of resource element (RE) carrying information originating from higher layers. The following downlink physical channels are defined (see Figure 5.3): Physical Broadcast Channel (PBCH), Physical Downlink Shared Channel (PDSCH), and Physical Downlink Control Channel (PDCCH). The PDCCH is used to transmit the DCI to the UE. The DCI is used to schedule DL transmissions on PDSCH and UL transmissions on PUSCH. The DCI includes (among other fields):

- Downlink assignments containing at least modulation and coding format, resource allocation, and hybrid-ARQ information related to DL-SCH;
- Uplink scheduling grants containing at least modulation and coding format, resource allocation, and hybrid-ARQ information related to UL-SCH.

The mappings between UL logical Channels, UL transport channels, and UL PHY channels are shown in Figure 5.4.

The following uplink physical channels are defined: Physical Random Access Channel (PRACH) used by UE to access the mobile network, Physical Uplink Shared Channel (PUSCH), and the Physical Uplink Control Channel (PUCCH). The PUCCH carries the UCI from the UE to the gNB. UCI consists of the following information: Channel State Information (CSI), HARQ acknowledgment (ACK/NAK), and Scheduling request. It is used by User Equipment (UE) to access the mobile network.

For details on NR channels refer to 3GPP specifications: TS 38.300 [4].

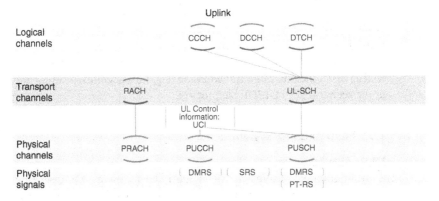

Figure 5.4 NR Uplink channels. Source: Adapted from [4].

5.2.8 NR Reference Signals

In addition to the physical channels, physical layer signals are defined in the Downlink and Uplink. A physical signal corresponds to a set of RE used by the physical layer but does not carry information originating from higher layers. These include Synchronization signals (PSS and SSS) or NR reference signals (RS). RS are sequences that are known to the receiver. They are tailored for specific roles and can be flexibly adapted for different deployment scenarios and spectrums.

The following physical signals are defined:

- Demodulation reference signals (DM-RS) for PBCH, PDSCH, PDCCH, PUSCH, and PUCCH: Designed for downlink/uplink channel estimation; coherent demodulation.
- Phase-tracking reference signals (PT-RS): It is a low-density pilot sequence sent at regular time intervals, used to enable tracking of phase noise in both UL/DL.
- Channel-state information reference signal (CSI-RS): UE-specific CSI-RS can be used for estimation of channel-state information (CSI) to further prepare feedback reporting to gNB to assist in MCS selection, beamforming, MIMO rank selection, and resource allocation.
- Tracking reference signals (TRS): Time and frequency TRS is designed for time/frequency tracking and estimation of delay/Doppler spread. TRS is configured as a CSI-RS with specific parameter restrictions (time/frequency location, RE pattern, etc.).
- Primary synchronization signal (PSS) and Secondary synchronization signal (SSS): Transmitted together with the PBCH. Each occupies 1 symbol

and 127 subcarriers. This pair of downlink signals is used by UE to find, synchronize to, and identify a network.

- Sounding reference signal (SRS): UE-specific SRS can be used for estimation of uplink CSI to assist uplink scheduling, and uplink power control, as well as assist the downlink transmission (e.g. the downlink beamforming in the scenario with UL/DL reciprocity).

5.2.9 Multi-antenna System

5.2.9.1 MIMO Schemes

The multi-antenna systems in NR support the following Multiple Input Multiple Output (MIMO) transmission schemes at both the UE and the gNB:

- Spatial multiplexing with DMRS-based closed loop, open-loop, and semi-open loop transmission schemes are supported. Both codebook and non-codebook–based transmission are supported in DL and UL.
- Spatial transmit diversity is supported by using specification transparent diversity schemes.
- Hybrid beamforming, including both digital and analog beamforming, is supported. Beam management with periodic and aperiodic beam refinement is also supported.

In NR, spatial multiplexing is supported with the following options:

Single code word is supported for 1–4 layer transmissions and two code words are supported for 5–8 layer transmissions in DL. Only single code word is supported for 1–4 layer transmissions in UL

Both open- and closed-loop MIMO are supported in NR, where for demodulation of data, receiver does not require knowledge of the precoding matrix used at the transmitter. Dynamic switching between different transmission schemes is also supported

Both single-user and multi-user MIMO are supported. For the case of single-user MIMO transmissions, up to eight orthogonal DM-RS ports are supported in DL, and up to four orthogonal DM-RS ports are supported in UL. For multi-user MIMO up to 12 orthogonal DM-RS ports with up to 4 orthogonal ports per UE are supported.

NR supports coordinated multipoint transmission/reception, which could be used to implement different forms of cooperative multi-antenna (MIMO) transmission schemes.

The use of multi-antenna technology in NR is focused on two objectives. First objective is to ensure sufficient coverage for NR deployment in over-6 GHz spectrum where propagation loss over wireless channels is significantly higher than that of sub-6 GHz spectrum. The second objective is to achieve higher spectral efficiency, which is especially important for sub-6 GHz spectrum.

5.2.9.2 Beam Management

Beam Management is a set of procedures to assist UE in setting its reception and transmission beams. It is composed of four different operations: Beam Indication, Beam Measurements and Reporting, Beam Recovery, and Beam Tracking and Refinement. These procedures are periodically repeated to update the optimal transmitter and receiver beam pair over time.

Beam Indication is used to assist UE in setting its reception and transmission beam properly for the reception of DL and transmission of UL, respectively. Beam Measurements and Reporting include procedures for providing gNB knowledge about feasible DL and UL beams for the UE.

For a large number of antennas, beams are narrow and beam tracking can fail; therefore, beam recovery procedures have also been defined where a device can trigger a beam recovery procedure. This procedure is used for rapid link reconfiguration against sudden blockages, i.e., fast re-aligning of gNB and UE beams. Moreover, a cell may have multiple transmission points, each with beams, and the beam management procedures allow for device-transparent mobility for seamless handover between the beams of different points. Beam Tracking and Refinement is a set of procedures to refine gNB and UE side beams. Additionally, uplink-centric and reciprocity-based beam management is possible by utilizing uplink signals.

Beam Management is used in Initial Access and Connected State. In Initial Access "Beam Management" relies on implicit determinations of used DL and UL beams while in Connected State it relies on gNB-controlled means to determine and select used DL and UL beams.

For NTN, the Release-15 and Release-16 beam management and BWP operation are considered as baseline. One-beam per cell and multiple-beam per cell are supported in existing NR specifications and are baseline for NR NTN.

5.2.9.3 Polarization Signaling in NTN

Satellite systems may employ more advanced antennas to create multiple spot beams on the ground. To mitigate inter-cell interference, frequency reuse schemes (Frequency reuse factor, $FRF > 1$) may be used. Spatial Frequency reuse techniques improve the SINR but inherently limit the per-beam bandwidth and the system capacity. The traditional Frequency

Reuse-3 (FRF-3) scheme, for example, offers protection against inter-cell interference. However, only a third of the spectral resources are used within each cell. In S-band, for example, in case of full frequency reuse ($FRF = 1$), each cell is configured with a bandwidth of 30 MHz, while in case of FRF-3, each cell is configured with a bandwidth of 10 MHz as the system bandwidth is divided into the three neighboring cells to decrease interference.

To increase the per-beam bandwidth while ensuring good interference isolation between beams, polarization reuse scheme can be used together with frequency reuse as shown in Figure 5.5: neighboring cells may use different polarization modes, including Right hand and Left hand circular polarizations (RHCP and LHCP). The Release-16 SI on solutions for NR to support NTN concluded that it is beneficial to signal the polarization mode for NTN in certain scenarios. Particularly when the UE is capable of differentiating RHCP and LHCP with the circularly or linearly polarized antennas.

NR NTN Release-17 specifies, the indication of polarization type used at satellite side for DL transmission. And the indication of the polarization type used at satellite side for UL reception. Two RRC parameters are introduced for polarization signaling in NR NTN:

- ntn-PolarizationDL: To indicate polarization information for Downlink transmission on service link: including RHCP and LHCP and Linear polarization. The indicated polarization information refers to the polarization type used on the satellite side for DL transmission.
- ntn-PolarizationUL: To indicate polarization information for Uplink service link. If this parameter is not indicated while ntnPolarizationDL is indicated, UE assumes the same polarization for UL and DL. The indicated polarization information refers to the polarization type used on the satellite side for UL reception.

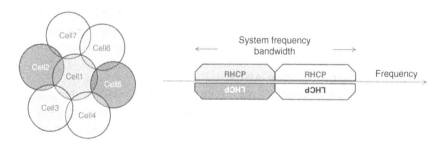

Figure 5.5 Polarization reuse scheme used together with frequency reuse.

Further, NTN Release-17 supports polarization signaling for target serving cells in handover command message. And polarization signaling for non-serving cells in RRM measurement configuration.

As Specified in 3GPP TS 38.306 [12], it is optional for UE to support the polarization signaling in NR NTN comprised of the following functional components:

- Support polarization indication reception in SIB indicating DL and/or UL polarization information using respective polarization type parameters to indicate: RHCP or LHCP or linear;
- Support polarization signaling for target serving cell in handover command message;
- Support polarization signaling for non-serving cells in RRM measurement configuration.

5.3 Downlink Synchronization Procedure in NTN

Cell search procedure is used by the UE to acquire time and frequency synchronization within a cell and detects the Cell ID of that cell. Both synchronization signals PSS and SSS are used by UE to acquire the downlink synchronization of the network, which comprises the following procedures: frequency synchronization, slot synchronization, frame synchronization, and cell ID detection.

The PSS is used for initial symbol boundary, CP, sub frame boundary, and initial frequency synchronization to the cell. The SSS is used for radio frame boundary identification. PSS and SSS together are used for the detection of physical layer Cell ID (PCI).

Synchronization signals (SS) and PBCH channel are packed as a single block that always moves together. They are jointly referred to as a SS block or SSB.

Other synchronization mechanisms are defined, e.g., for radio link monitoring, transmission timing adjustments (as described in Section 5.4.1), and timing for cell activation/deactivation.

To accommodate the large Doppler shift in LEO/MEO–based NTN Satellite systems, beam-specific pre-compensation of common frequency shift can be applied to the signal in the DL, which can be conducted with respect to a reference point (RP) (e.g., the spot beam center) such that the DL signal received at this RP is appeared to have zero Doppler shift from the satellite movement. This solution was a potential enhancement extensively discussed during the study and normative phases of

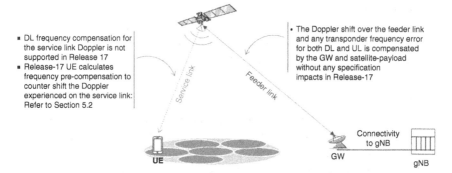

Figure 5.6 DL frequency compensation for Doppler in Release-17. Source: Adapted from [13].

NR NTN support. It is observed in [2] via simulations that for DL initial synchronization, with common frequency offset compensation on the DL service link, robust performance can be provided by the SSB design in Rel-15/16 in case of GEO and LEO. However, the DL initial synchronization is also possible on NGSO radio link without pre-compensation of the frequency offset and no issues have been identified based on Rel-15/16 NR design. Only additional complexity is needed at UE receiver to achieve robust DL initial synchronization performance based on Rel-15 SSB as captured in [2].

Therefore, NR NTN SI concluded that no further enhancement on the SSB is needed to support NTN. Moreover, during the NTN normative phase, it was agreed that DL frequency compensation by gNB for the service link Doppler is not supported in Release-17. Further, as illustrated in Figure 5.6, the Doppler shift over the feeder link and any transponder frequency error for both Downlink and Uplink is compensated by the NTN GW and satellite-payload without any specification impacts in Release-17.

5.4 Uplink Synchronization Procedure in NTN

5.4.1 Uplink Timing Control

The propagation delay over an NTN radio link is much longer than in terrestrial cellular networks, ranging from several milliseconds to hundreds of milliseconds, depending on the altitude and the type of the NTN platform. This delay largely exceeds the TTI of NR which is equal to or less than 1 ms. To accommodate such a long propagation delay, 3GPP Release-17 specified many enhancements on timing aspects in NR from physical layer to higher layers, including the timing advance (TA) mechanism.

5.4.1.1 Uplink Timing Control in NR TN

In NR, uplink transmission from all UEs within a serving cell shall be synchronized when received at the Base Station reception point: Inter-symbol interference shall be avoided and thereby all uplink transmissions shall be received with a time spread, which is less than the duration of the CP. The requirements for UE transmit timing and TA adjustment accuracy are specified in clause 7 of 3GPP TS 38.133 [14].

The uplink timing of each individual UE is controlled by the Base Station with MAC message as defined in clause 5.2 of TS 38.321. This is mainly applicable to the PUSCH, PUCCH, and SRS. In NR TN, the PRACH can only use a fixed timing offset because the UE transmits the PRACH preambles before the gNB is able to provide any timing adjustment. The TA refers to the negative offset between the start of a given DL frame as observed by the UE and the start of the corresponding frame in the UL. Uplink/Downlink Radio frame timing at the UE is shown in Figure 5.7.

The TA applied by the UE is given by the following formula:

$$T_{TA} = (N_{TA} + N_{TA,offset}) \times T_c$$
$$Tc = 1/(480\,000 \times 4096)$$

For initial access, random-access procedure is used in order for UE to achieve Uplink synchronization and obtain the resource for RRC Connection Request message. Initial access procedure is triggered by the UE by

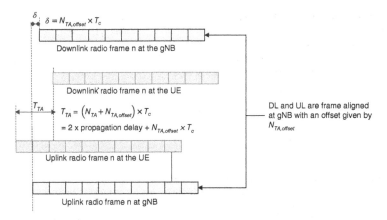

Figure 5.7 Uplink/Downlink radio frame timing in TN.

transmitting a random access PRACH preamble (Msg1[3] or MSGA[4]) with $N_{TA} = 0$, then the UE receives its first Timing Advance Command (TAC) within the Random Access Response (RAR) message (a.k.a MSG2). Within the TAC, a set of 12 bits is used to indicate a T_A value from 0 to 3846. The UE uses this value to calculate N_{TA} as follows:

$$N_{TA} = T_A \times 16 \times \frac{64}{2^\mu}$$

where μ is the SCS index.[5]

At PRACH preamble transmission $N_{TA} = 0$. $N_{TA,offset}$ is provided in SIB1 and it was specified to ensure that a UL radio frame finishes before the start of the subsequent DL radio frame, which is useful for a TDD Base Station. But this offset can be also used to avoid gNB to gNB interference. A UE can be provided a value $N_{TA,offset}$ of a TA offset for a serving cell by n-TimingAdvanceOffset. If the UE is not provided n-TimingAdvanceOffset for a serving cell, the UE determines a default value $N_{TA,offset}$ of the TA offset for the serving cell as described in TS 38.133 [14]. $N_{TA,offset}$ has unit of Tc. The default value of $N_{TA,offset}$ is set as 25 600 for FR1 band. In FR2, there is only a unique supported value of $N_{TA,offset}$, thereby, it may not be signaled. The values of $N_{TA,offset}$ are depicted in Table 5.2.

The maximum TA in NR TN which can be compensated during initial access is calculated in Table 5.3.

In RRC_CONNECTED, the gNB is responsible for maintaining the TA to keep Layer 1 synchronized [4]. TA updates are signaled by the gNB to the UE via MAC CE commands as defined in clause 5.2 of TS 38.321. a

Table 5.2 The value of $N_{TA,offset}$.

Frequency range and band of cell used for uplink transmission	$N_{TA,offset}$ (Unit: T_c)	$N_{TA,\,offset} \times T_c$
FR1 FDD band without LTE-NR coexistence case or FR1 TDD band without LTE-NR coexistence case	25 600	13 µs
FR1 FDD band with LTE-NR coexistence case	0	0 µs
FR1 TDD band with LTE-NR coexistence case	39 936	20 µs
FR2	13 792	7 µs

---

3 Preamble transmission of the random-access procedure for 4-step RA type.
4 Preamble and payload transmissions of the random-access procedure for 2-step RA type.
5 Multiple SCS are supported including 15 kHz ($\mu = 0$), 30 kHz ($\mu = 1$), 60 kHz ($\mu = 2$), and 120 kHz ($\mu = 3$). Note that $SCS = 2^\mu$ 15 kHz ($\mu = 0..3$).

Table 5.3 Maximum timing advance update per TAC in RAR.

Numerology (μ)	0	1	2	3
SCS = 15 * 2^μ kHz	15	30	60	120
Maximum timing advance update during initial access in NR TN	+2 ms	+1 ms	+0.5 ms	+0.25 ms
Maximum cell range calculated from the maximum TA	300 km	150 km	75 km	37.5 km

Source : Dahlman et al. [18]/with permission of Elsevier.

Table 5.4 Maximum timing advance update per TAC in MAC CE.

Numerology (μ)	0	1	2	3
SCS = 15 * 2^μ kHz	15	30	60	120
Maximum timing advance compensated via timing advance command in MAC CE (ms)	±0.017	±0.008	±0.004	±0.002

TA command, TA, for a TAG indicates adjustment of a current N_{TA} value, N_{TA_old}, to the new N_{TA} value, N_{TA_new}, by index values of $T_A = 0, 1, 2, ..., 63$, where for a SCS of 2^μ. 15 kHz, $N_{TA\_new} = N_{TA\_old} + (T_A - 31).16.\frac{64}{2^\mu}$ [4].

The maximum TA update per TAC in MAC CE is shown in Table 5.4.

Such commands restart a specific timer, which indicates whether the Physical Layer can be synchronized or not: when the timer is running, it is considered synchronized; otherwise, it is considered non-synchronized in which case uplink transmission can only take place through MSG1/ MSGA [4].

5.4.1.2 Uplink Timing Control in NR NTN

Considering the large RTD in satellite radio link, e.g., up to 541.46 ms for GEO and 25.77 ms for LEO 600 km, it is clear that the value of TA in the TAC during initial access as shown in Table 5.3 is far from sufficient. Further, the high delay drift in case of non-GEO satellites can be a problem for TA control in Connected state. Indeed, due to a high radial velocity between the satellite and the UE the propagation delay changes quickly: In case of LEO transparent payload, for instance, the maximum delay variation as seen by the UE can be up to ±40 μs/s. And thereby, if only closed-loop timing control is used as in TN, the maintenance of the TA during connected state may become rather challenging because the user-specific timing adjustment commands (TAC in MAC CE) must be sent frequently leading to critical DL signaling overhead. Also, increasing the size of MAC-CE timing control command alone does not solve the problem: Assuming a one-way delay of 12.88 ms, a timing control command sent by the Network that is accurate at the time of its transmission can be off by 0.51 μs at the time of its arrival, which is larger than the 10% of CP duration for 15 kHz SCS, 4.76 μs.

Solutions have been specified in 3GPP Release-17 to meet the challenges related to large round trip delay (RTD) and timing drift: NTN UE shall support the uplink time pre-compensation (described in this section) and frequency pre-compensation (refer to Section 5.4.2), timing relationship enhancements (refer to Section 5.5) and a set of other NTN essential features, e.g., timers extension in MAC/RLC/PDCP layers and RACH adaptation to handle long RTT and introduction of several higher layer parameters and NTN specific SIB (i.e. SIB19) to assist the UE in uplink time and frequency synchronization. NTN UE may support these NTN features in GSO scenario or NGSO scenario or in both scenarios.

For uplink time synchronization in NR NTN access, the UE should perform the uplink time pre-compensation. To this aim, the UE is assisted by its GNSS and by the network. The network broadcasts assistance information in SIB19[6] in the serving NTN cell, including ephemeris information and higher layer Common-TA-related parameters. The UE shall acquire a valid GNSS[7] position as well as assistance information before and during connection to an NTN cell. Then, the UE calculates UE-specific TA corresponding to the round-trip time or RTT[8] on the service link, based on its GNSS-acquired position and the serving satellite ephemeris. If Common-TA-related parameters are indicated, the UE calculates common TA corresponding to any common delay, e.g., RTT on the feeder link, according to the parameters provided by the network. UE considers common TA as 0 if these parameters are not provided. The UE performs the pre-compensation of the calculated TA in its uplink transmissions,

---

6 SIB19 is a new System Information Block introduced for NTN. It contains NTN-specific parameters for serving cell and/or neighbor cells as defined in TS 38.331 [15].

7 A UE shall set **gnss-Location-r16** field within UE capability parameters to supported if it indicates the support of **nonTerrestrialNetwork-r17**. This field indicates whether the UE is equipped with a GNSS or A-GNSS receiver that may be used to provide detailed location information along with SON, MDT, and NTN-related measurements in RRC_CONNECTED, RRC_IDLE, and RRC_INACTIVE state.

8 In essence, in TN the RTT corresponds to $2 \times$ *one way propgation delay*. In NTN $RTT =$ *one way propgation delay*$_{TX}$ + *one way propgation delay*$_{RX}$. In fact, depending on the orbit, the propagation delay at transmission (*one way propgation delay*$_{TX}$) may be different to the propagation delay at reception (*one way propgation delay*$_{RX}$) because of satellite movement.

including PRACH preamble transmissions and uplink transmissions during the RRC_CONNECTED state.

For TA update in RRC_CONNECTED state, combination of both open-loop (i.e., UE autonomous TA estimation, and common TA estimation) and closed-loop (i.e. received TA commands in RAR message or MAC CE) is supported.

3GPP Release-17 specified the following formula for TA calculation that shall be applied by NTN Ues for PRACH preamble transmission and in RRC_CONNECTED state:

$$T_{TA} = \left(N_{TA} + N_{TA,offset} + N_{TA,adj}^{common} + N_{TA,adj}^{UE} \right) \times T_c$$

where:

- N_{TA} and $N_{TA,offset}$ are specified in 3GPP TS 38.213 [6] and 3GPP TS 38.211 [5] as part of the existing TA Control (refer to Section 5.4.1.1)
- $N_{TA,adj}^{common}$ is a network-controlled common TA, and may include any timing offset deemed necessary by the network (e.g. feeder link delay). It is derived from the higher-layer parameters **ta-Common**, **ta-CommonDrift**, and **ta-CommonDriftVariation** if configured, otherwise $N_{TA,adj}^{common} = 0$
- $N_{TA,adj}^{UE}$ is UE self-estimated TA to pre-compensate for the service link delay. It is computed by the UE based on UE position and serving satellite-ephemeris-related higher-layers parameters if configured, otherwise $N_{TA,adj}^{UE} = 0$
- T_c is the NR basic time unit: $T_c = \frac{1}{480\,000 \times 4096}$ [5] Uplink and Downlink radio frame timing at the gNB and at the UE in NTN is illustrated in (Figure 5.8).

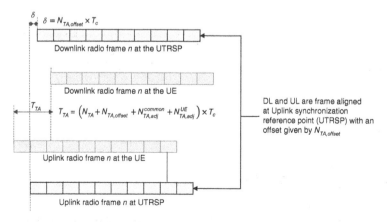

Figure 5.8 Uplink/Downlink radio frame timing at the gNB and the UE in NTN.

In the above formula, N_{TA} is updated via closed-loop TA control. The same principle as in TN is used. That is, N_{TA} update based on TA Command field in msg2[9]/msgB[10] and MAC CE TA command is used for UL timing alignment correction: When TAC (T_A) in msg2/msgB is received, UE receives the first adjustment and N_{TA} is updated as: $N_{TA} = T_A \cdot 16 \cdot \frac{64}{2^\mu}$, where, T_A is the TAC field in msg2/msgB. When TAC (T_A) provided within the MAC CE is received, N_{TA} is updated as follows: $N_{TA\_new} = N_{TA\_old} + (T_A - 31) \cdot \frac{16.64}{2^\mu}$, where, T_A is the TAC field received in MAC CE command.

$N_{TA,adj}^{UE}$ and $N_{TA,adj}^{common}$ are autonomously updated by the UE (i.e. open-loop TA control) as specified in TS 38.213 [6]: Using higher-layer ephemeris parameters for the serving satellite, the UE calculates $N_{TA,adj}^{UE}$, using serving satellite position and its own position, to pre-compensate the two-way transmission delay on the service link. Note that Release-17 does not specify how the UE calculates and updates $N_{TA,adj}^{UE}$. This UE-specific TA is therefore left for UE implementation.

When the reference point for uplink time synchronization (UTSRP) is not at the satellite, the UE needs to compensate the feeder link's RTD and any timing offset considered necessary by the network. Hence, the timing drift on the feeder link shall be compensated by the UE with sufficient accuracy.

To pre-compensate the two-way transmission delay between the UTSRP and the satellite, $N_{TA,adj}^{common}$ is derived by the UE based on $Delay_{common}(t)$, which can be obtained as:

$$Delay_{common}(t) = \frac{TA_{Common}}{2} + \frac{TA_{CommonDrift}}{2} \times (t - t_{epoch})$$
$$+ \frac{TA_{CommonDriftVariant}}{2} \times (t - t_{epoch})^2$$

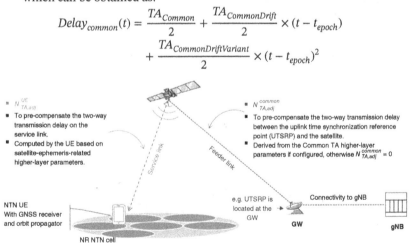

- $N_{TA,adj}^{UE}$
- To pre-compensate the two-way transmission delay on the service link.
- Computed by the UE based on satellite-ephemeris-related higher-layer parameters.

- $N_{TA,adj}^{common}$
- To pre-compensate the two-way transmission delay between the uplink time synchronization reference point (UTSRP) and the satellite.
- Derived from the Common TA higher-layer parameters if configured, otherwise $N_{TA,adj}^{common} = 0$

NTN UE
With GNSS receiver
and orbit propagator

NR NTN cell

e.g. UTSRP is located at the GW

Connectivity to gNB

GW

gNB

Figure 5.9 UE-specific TA and Common TA.

9 Random-Access Response message.
10 Response to MSGA in the two-step random-access procedure. MSGB may consist of response(s) for contention resolution, fallback indication(s), and backoff indication.

where t_{epoch} is the epoch time of the higher-layer parameters TACommon, TACommonDrift, and TACommonDriftVariation.

This $Delay_{common}(t)$ gives the distance at time t between the satellite and the UTSRP divided by the speed of light. The UTSRP is the point at which DL and UL are frame-aligned with an offset given by $N_{TA,offset}$ (Figure 5.9).

The Uplink Time Synchronization Reference Point (UTSRP): The uplink time synchronization reference point (UTSRP) is defined as the point at which DL and UL are frame-aligned with an offset given by $N_{TA,offset}$.

UTSRP can be located at the Base Station (as for TN), onboard the satellite, or possibly a RP on the feeder link. Based on UTSRP position, the Base Station may indicate the higher layer common TA parameters that are deemed necessary.

In essence, the UTSRP position is only known to the network. The common TA parameters would be determined and broadcast by the network and would implicitly define the RP. The exact location of the RP would be an internal matter to the network. That is, the UE does not need to know the exact location of the UTSRP.

If the UTSRP is within satellite payload, the Base Station will not indicate the common TA parameters: the UE needs to compensate only two-way delay on the service link and consider $N_{TA,adj}^{common} = 0$. This may be the case if the Base Station or part of it (e.g. Radio Unit or Distributed Unit) is onboard the satellite.

If the UTSRP is not located on the satellite, the network will indicate the common TA parameters that will be used by the UE to derive the two-way transmission delay between the UTSRP and the satellite. This may be the case if the UTSRP is located on the NTN GW: Thereby, the common TA parameters will be used to derive the two-way transmission delay on the feeder link. Note that in this case, the UE may be able to determine the GW location (based on triangulation positioning method using broadcast Common TA parameters at different instants). If the GW location should not be disclosed, the network may indicate appropriate Common TA parameters putting the UTSRP away from the actual GW location (e.g. a point on the feeder link). Note that the UTSRP cannot be put on the service link because only positive Common TA values are supported in Release-17.

As part of UE capability parameters, the field **uplinkPreCompensation-r17** is used to indicate whether the UE supports the uplink time and frequency pre-compensation and timing relationship enhancements. As specified in 3GPP TS 38.306 [12], support of uplink time pre-compensation in NTN bands is mandatory for UE supporting NR NTN access. The NTN UE shall support the uplink time pre-compensation that comprises the following functional components:

- Support of UE-specific TA calculation based on its GNSS-acquired position and the serving satellite ephemeris
- Support of common TA calculation according to the parameters provided by the network (UE considers common TA as 0 if the parameters are not provided)
- For TA update in RRC_CONNECTED state, support of combination of both open (i.e. UE autonomous TA estimation, and common TA estimation) and closed (i.e. received TA commands) control loops
- Support of pre-compensation of the calculated TA in its uplink transmissions

UE supporting NR NTN access shall satisfy the uplink timing requirements for initial access (i.e., PRACH transmission) and for uplink transmissions in RRC Connected state, including the UE initial transmission timing error requirement for RRC_CONNECTED, gradual timing adjustment requirement, and TA adjustment accuracy requirement. These requirements are specified in 3GPP TS 38.133 [14] and captured in Chapter 7.

5.4.1.3 NTN Higher-Layer Parameters for Uplink Timing Control

The satellite ephemeris information, common TA-related parameters, validity duration for UL sync information, and epoch time are signaled to the UE within **ntn-Config** Information Element (IE) as part of network assistance information needed for the UE to access NR via NTN access. These parameters are used by the UE to calculate $N_{TA,adj}^{UE}$ and $N_{TA,adj}^{common}$ as discussed in Section 5.4.1.2. To keep track of satellite movement, the Base Station needs to update regularly NTN Higher-Layer parameters. Thereby, related fields broadcast in SIB19 are excluded when determining changes in SI, i.e., changes to these parameters should neither result in SI change notifications nor the modification of **valueTag**[11] in SIB1.

Satellite Ephemeris Information Two satellite ephemeris formats are supported in 3GPP Release-17: Position and velocity (PV) state vector in Earth-Centered Earth-Fixed (ECEF) indicated in 17 bytes payload and

---

11 **valueTag** is signaled in SIB1. It is an integer value (0–31) used to indicate that the content of a specific SIB has changed, and the UE should reacquire the SIB: It is incremented if the content of the SIB has changed. NTN network assistance information in SIB19 is an exception, which allow SIB content to be modified without a modification of **valueTag**. There are other exceptions in existing specifications such as the timing information within SIB9, which can be updated without impacting **valueTag**.

orbital parameters in Earth-centered intertial (ECI) indicated in 21 bytes payload. Satellite ephemeris data is provided within **EphemerisInfo** in **ntn-Config** IE. Satellite ephemeris parameters specified in Release-17 are listed in Table 5.5.

Satellite ephemeris is part of the NTN-related parameters that shall be provided by O&M to the gNB providing NTN access as listed in clause 16.14.7 of 3GPP TS 38.300 [4]: The measurements of the satellite PV are typically GNSS-based measurements performed on-board. These measurements are then collected in the NTN Control Center (NCC) where the satellite orbit determination (OD) is performed as illustrated in Figure 5.10. The satellite OD can be more or less complex depending on the models considered, the quantity of measurements available and the algorithms used. After OD, the orbit prediction is performed at NCC and provided to the Base Station via O&M interface. The orbit prediction accuracy depends on several factors: The accuracy of the OD used to derive the satellite ephemeris, the accuracy of the orbit propagation model, and the time horizon over which the prediction is made: The initial errors combined with the propagation model errors tend to increase the prediction errors when the prediction period increases. Thereby, the NCC needs to periodically provide the Base Station with updated ephemeris information depending on the required ephemeris precision.

3GPP Release-17 has defined requirements for the UE timing error, i.e., Te_NTN in NTN (refer to Chapter 7). The Te_NTN includes Te_GNSS and Te_SAT. Te_GNSS is the maximum RTT error between UE and satellite due to UE GNSS position estimation error and Te_SAT is the error from the orbital propagator model used at UE side to estimate the satellite position. Te-SAT should integrate the following contributions (i)–(iii): (i) serving-satellite position estimation error due to OD and prediction at NCC/gNB side. (ii) quantization error linked to bit allocation used for serving satellite ephemeris format. (iii) Serving-satellite position estimation error due to orbit propagation at UE side. Clearly, (iii) is dependent on (i) and (ii).

Serving-satellite position estimation error due to orbit propagation at UE side is evaluated in [16] via simulation for an orbit LEO 600 km, near circular, inclination 53° based on a typical Precision Orbit Determination (POD); which leads to an initial 3D position RMS error of 0.5 m and 3D velocity RMS error of 0.5 mm/s, which allows satellite position prediction at NCC/Base Station 5 minutes ahead with a maximum error of 3.87 m [17]. 3GPP Release-17 does not specify a specific orbital propagator model to be used on UE side; this is left to implementation. Three propagation models in UE were evaluated: NUM6x6, NUM2x2, and KEPLER.

Table 5.5 Supported Satellite ephemeris parameters in 3GPP Release-17.

Parameter name	Description
	Satellite ephemeris parameters based on state vector:
positionX positionY positionZ	Indicate the x,y,z-coordinate of serving satellite position state vector in ECEF. The unit is m. Value range: $-33554432...33554431$ The quantization step is 1.3 m for position: Actual value = field value * 1.3
velocityVX velocityVY velocityVZ	Indicate the x,y,z-coordinate of serving satellite velocity state vector in ECEF. The unit is m/s. Value range: $-131072...131071$ The quantization step is 0.06 m/s for Velocity: Actual value = field value * 0.06.
	Satellite ephemeris parameters based on orbital elements (Keplerian):
semiMajorAxis	Indicate the semi-major axis α. The unit is m. Value range: $650\,000\,0..430\,000\,00$ The quantization step is 4.249×10^{-3} m: Actual value = $650\,000\,0 +$ field value * $(4.249 * 10^{-3})$.
eccentricity	Indicate eccentricity e. Value range: $0...0.015$ The quantization step is 1.431×10^{-8}: Actual value = field value * $(1.431 * 10^{-8})$
periapsis	Indicate the argument of periapsis ω. The unit is Radian. Value range: $0...268\,435\,455$ The quantization step is 2.341×10^{-8} rad: Actual value = field value * $(2.341 * 10^{-8})$.
longitude	Indicate the longitude of ascending node Ω. The unit is Radian. Value range: $0...268\,435\,455$ $(0...2\pi)$ The quantization step is 2.341×10^{-8} rad: Actual value = field value * $(2.341 * 10^{-8})$.
inclination	Indicate the inclination i. The unit is Radian. Value range: $-671\,088\,64...671\,088\,63$ $(-\pi/2...+\pi/2)$ The quantization step is 2.341×10^{-8} rad: Actual value = field value * $(2.341 * 10^{-8})$.

(Continued)

Table 5.5 (Continued)

Parameter name	Description
meanAnomaly	Indicate the mean anomaly M at epoch time. The unit is Radian. Value range: $0\ldots268\,435\,455\ (0\ldots2\pi)$ The quantization step is 2.341×10^{-8} rad: Actual value = field value * $(2.341 * 10^{-8})$.

Source: Adapted from [13].

The broadcast serving satellite ephemeris is in state vector PV ephemeris format. According to the simulation results depicted in Table 5.6, the UE can predict the satellite PV errors with sufficient accuracy: UE can predict 60 seconds ahead the satellite position with less than 10 m error. And 300 seconds ahead with less than 39 m position error when the orbit propagator used at the UE is NUM2x2 or NUM6x6.

Common TA Higher-layer Parameters   Higher-layer parameters **ta-Common**, **ta-CommonDrift**, **ta-CommonDriftVariation** are signaled to the UE as part of the NTN assistance information in SIB19 in 26 bits,19 bits, and 15 bits respectively. They are used by the UE to calculate the one-way transmission delay between the UTSRP and the satellite given by the following second-order approximation:

$$Delay_{common}(t) = \frac{TA_{Common}}{2} + \frac{TA_{CommonDrift}}{2} \times (t - t_{epoch})$$
$$+ \frac{TA_{CommonDriftVariant}}{2} \times (t - t_{epoch})^2$$

In essence, the TA applied by the UE should compensate for the delay in the UL signal path from UE to the UTSRP at the time of transmission of a given UL slot and the delay in the DL signal path from UTSRP to UE of the corresponding DL slot. Therefore, for a given uplink slot n, the common TA i.e. $N_{TA,adj}^{common}$ can be derived by the UE as:

$Delay_{common}$(time of transmission of UL slot n)

$+ Delay_{common}$(corresponding DL slot n)

Further, it was agreed in 3GPP to signal Common TA parameters corresponding to the two-way RTT to the UE from the UTSRP. Therefore, these parameters are divided par a factor of 2 in the formula of $Delay_{common}(t)$.

The definitions of higher-layer Common TA parameters are given in Table 5.7.

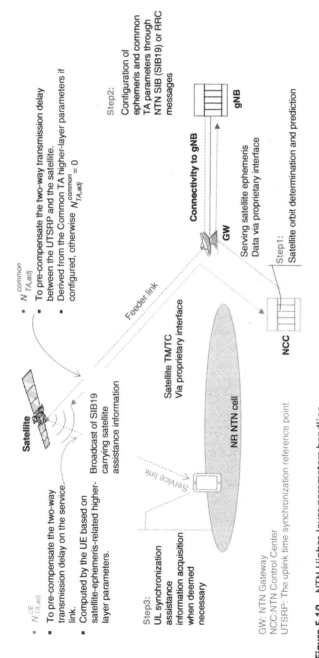

- $N_{TA,adj}^{UE}$
  - To pre-compensate the two-way transmission delay on the service link.
  - Computed by the UE based on satellite-ephemeris-related higher-layer parameters.

Broadcast of SIB19 carrying satellite assistance information

- $N_{TA,adj}^{common}$
  - To pre-compensate the two-way transmission delay between the UTSRP and the satellite.
  - Derived from the Common TA higher-layer parameters if configured, otherwise $N_{TA,adj}^{common} = 0$

Satellite

Step2:
Configuration of ephemeris and common TA parameters through NTN SIB (SIB19) or RRC messages

Connectivity to gNB

gNB

Serving satellite ephemeris
Data via proprietary interface

GW

Step1:
Satellite orbit determination and prediction

Feeder link

Satellite TM/TC
Via proprietary interface

NCC

Step3:
UL synchronization assistance information acquisition when deemed necessary

Service link

NR NTN cell

GW: NTN Gateway
NCC:NTN Control Center
UTSRP: The uplink time synchronization reference point

Figure 5.10 NTN Higher-layer parameters handling.

Table 5.6 Orbit propagation errors at the UE [16].

PROPAGATION MODEL in UE:	NUM6x6						
	Ephemeris format: PV						
Propagation duration (s)	5	10	15	20	30	60	300
Satellite position error at the UE (m) — Mean	2.95	3.09	3.25	3.41	3.79	5.19	19.11
Sigma	0.82	0.90	0.98	1.06	1.22	1.69	5.80
Max (99.7%)	**4.77**	**5.07**	**5.49**	**5.91**	**6.59**	**9.29**	**32.90**
Satellite velocity error at the UE (mm/s) — Mean	59.75	59.82	59.89	59.97	60.12	60.59	64.68
Sigma	18.26	18.26	18.26	18.26	18.27	18.27	19.11
Max (99.7%)	99.32	99.38	99.46	99.54	99.64	99.71	113.93

PROPAGATION MODEL in UE:	NUM2x2						
	Ephemeris format: PV						
Propagation duration (s)	5	10	15	20	30	60	300
Satellite position error at the UE (m) — Mean	2.95	3.09	3.25	3.42	3.79	5.20	19.94
Sigma	0.82	0.90	0.98	1.06	1.23	1.70	6.74
Max (99.7%)	**4.77**	**5.07**	**5.49**	**5.92**	**6.63**	**9.45**	**38.96**
Satellite velocity error at the UE (mm/s) — Mean	59.87	60.06	60.25	60.45	60.86	62.18	75.70
Sigma	18.26	18.26	18.26	18.27	18.29	18.46	25.24
Max (99.7%)	99.40	99.62	99.83	100.54	101.99	106.63	159.19

PROPAGATION MODEL in UE:	KEPLER						
	Ephemeris format: PV						
Propagation duration (s)	5	10	15	20	30	60	300
Satellite position error at the UE (m) — Mean	2.96	3.16	3.52	4.13	6.42	22.32	544.41
Sigma	0.82	0.91	1.06	1.31	2.03	4.38	79.40
Max (99.7%)	**4.80**	**5.35**	**6.23**	**7.63**	**11.75**	**33.87**	**721.73**
Satellite velocity error at the UE (mm/s) — Mean	86.02	132.50	189.20	248.80	369.85	736.28	3608.91
Sigma	24.53	35.90	43.07	48.94	62.32	109.16	528.60
Max (99.7%)	156.46	223.52	299.80	375.48	528.21	994.36	4741.30

Epoch Time of Assistance Information Satellite ephemeris and Common TA parameters have the same epoch time referred to as **epochTime** in 3GPP TS 38.331 [15]. This **epochTime** defines the time reference for satellite ephemeris and common TA. In essence, it also indicates the start time of the validity duration configured by the network for NTN-assistance information (refer to Section Uplink Synchronization Validity Duration).

Table 5.7 Common TA parameters definition.

Parameter name	Description
ta-Common	TACommon is a network-controlled common TA, and may include any timing offset considered necessary by the network.
	TACommon with value of 0 is supported. The granularity of TACommon is 4.072×10^{-3} μs
	Values are given in units of corresponding granularity
	Value range: 0...664 857 57 (i.e.: 0...270.73 ms)
ta-CommonDrift	Indicate drift rate of the common TA
	The granularity of TACommonDrift is 0.2×10^{-3} μs/s
	Values are given in unit of corresponding granularity
	Value range: −257 303... + 257 303 (i.e: −52.387 μs/s ... + 52.387 μs/s)
ta-CommonDriftVariation	Indicate drift rate variation of the common TA
	The granularity of TACommonDriftVariation is 0.2×10^{-4} μs/s^2
	Values are given in unit of corresponding granularity
	Value range: 0...289 49 (i.e.: 0...0.5894 μs/s^2)

The **epochTime** can be explicitly provided through SIB19, or through dedicated signaling. When explicitly indicated, **epochTime** is the starting time of a DL sub-frame, indicated by a SFN and a sub-frame number signaled together with the assistance information as illustrated in Figures 5.11 and 5.12.

As the SFN period equals 1024, i.e., the SFN repeats itself after 1024 frames or 10.24 seconds, there might be an ambiguity in the interpretation of SFN indicating epoch time. 3GPP Release-17 resolved this ambiguity in a different way for serving and neighbor cell as follows:

- For serving cell, the field SFN if not indicating the current SFN, it will point to a future SFN, specifically the next upcoming SFN after the frame where the message indicating the **epochTime** is received. As can be seen from the example in Figure 5.11, the SIB19 including NTN-assistance information is transmitted in sfn7 and repeated in sfn15 and in both

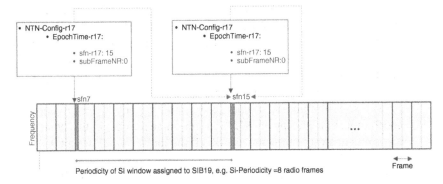

Figure 5.11 Epoch time explicitly provided in *ntn-Config*-r17 for serving cell.

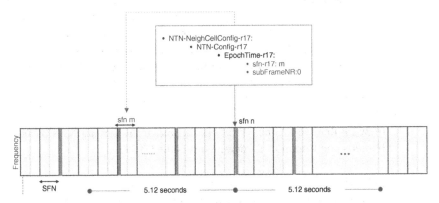

Figure 5.12 Epoch time explicitly provided in *ntn-Config*-r17 for neighbor cell (*ntn-Config* provided via NTN-NeighCellConfig).

transmissions, the **epochTime-17** is pointing to the starting time of sub-frame 0 of sfn 15 {sfn − r17 = 15, subFrameNR = 0}.

- For neighbor cell, the SFN indicates the SFN nearest to the frame where the message indicating the **epochTime** is received. With this clarification, the behavior of the UE is clear when interpreting the SFN indicating epoch time as shown in Figure 5.12: When receiving the SIB19 at SFN *n* with **EpochTime-r17** pointing to sfn *m*, there is a unique sfn *m* nearest to sfn *n* within the time window centered in sfn *n*.

The network may not explicitly indicate the **epochTime** through SIB. And thereby the **epochTime** filed may not be present within **ntn-Config** in SIB19. In this case, the time reference for assistance information is implicitly known by the UE as the end of the SI window where the SIB 19 is transmitted.

If this field is absent in **ntn-Config** provided via **NTN-NeighCellConfig**, the UE uses epoch time of the serving cell, otherwise the field is based on the timing of the serving cell, i.e., the SFN and sub-frame number indicated in this field refers to the SFN and sub-frame of the serving cell. In case of handover or conditional handover, this field is based on the timing of the target cell, i.e., the SFN and sub-frame number indicated in this field refers to the SFN and sub-frame of the target cell. For the target cell, the UE considers epoch time, indicated by the SFN and sub-frame number in this field, to be the frame nearest to the frame in which the message indicating the epoch time is received.

Further, the definition of a reference point (RP) for the **epochTime** is also needed. Indeed, the UE and the Base Station shall share the same understanding about when the starting time of a DL sub-frame (or the end of the SI window where the SIB 19 is scheduled) that defines the epoch time is passing via the RP: To take into account the aging of assistance information; that is, if this RP is set to be at Base Station, the UE needs to take into account the aging of this **epochTime** by a duration equal to one-way UE-gNB delay. Otherwise, the UE would initialize its propagator with an error on satellite position up to 97.45 m[12] in case of LEO600km orbit with a transparent payload.

In 3GPP Release-17, the RP for epoch time of the serving NTN payload ephemeris and Common TA parameters is the UTSRP.

Uplink Synchronization Validity Duration For UE-specific TA updates, the UE needs to acquire satellite ephemeris and use its own orbit propagator to predict its specific TA for a certain duration without the need of acquiring new ephemeris information. The newly acquired ephemeris will be valid only during a time window depending on serving-satellite position estimation error due to OD and prediction by the network, serving-satellite position estimation error due to orbit propagation at UE, quantization error, and the maximum tolerable transmission timing error. As shown within the simulations results in Table 5.6, a UE with NUM6x6 orbit propagator can predict the satellite position 60 seconds ahead with less than 9.29 m error, which corresponds to 0.03 μs or 0.65% of the CP in case of 15 kHz SCS. With a propagation duration of 300 seconds, the satellite position

---

12 In case of a transparent payload, the one-way delay including the delay on both service and feeder links is equal to 12.89 ms. If the UE considers the satellite position is corresponding to the time of arrival ($t0$) of ephemeris information and not the time when ephemeris information was transmitted from the Base Station ($t0 - 12.89$ ms), as the satellite speed is equal to 7.56 m/ms, during the propagation delay of 12.89 ms the satellite will move by approximately 97.45 m.

error at the UE will be equal to 32.90 m, which corresponds to 0.1 μs or 2.3% of the CP in case of 15 kHz SCS. Similarly, for Common TA update, the Common TA parameters acquired in SIB19 will be valid only during a validity time window, which depends on the Common TA parameters used for Common TA estimation[13] and the maximum tolerable error on common TA estimation. Therefore, a time alignment validity timer should be introduced and associated with open-loop TA control.

In essence, satellite ephemeris and common TA parameters may have different validity durations. Indeed, using highly sophisticated orbit propagator at the network and UE may allow longer validity duration for satellite ephemeris information. Whereas for common TA calculation, only a second-order approximation is adopted in Release-17, which may not allow longer validity duration as for ephemeris. However, because satellite ephemeris and common TA parameters are signaled within the same SIB, 3GPP Release-17 defined a single validity duration, i.e., **ntn-UlSyncValidityDuration** for both serving satellite ephemeris and common TA related parameters (refer to Table 5.8). It is configured by the network to indicate the maximum time during which the UE can apply assistance information (i.e. serving and/or neighbor satellite ephemeris and Common TA parameters) without having acquired new assistance information. **ntn-UlSyncValidityDuration** is configured per cell and indicated in four bits with a value from the set {5s, 10s, 15s, 20s, 25s, 30s, 35s, 40s, 45s, 50s, 55s, 60s, 120s, 180s, 240s, 900s}. Validity duration of 900 seconds could be configured for a GEO-based NTN access. Serving and neighbor cells may have different **ntn-UlSyncValidityDuration**. If this field is absent in **ntn-Config** provided via **NTN-NeighCellConfig**, the UE uses validity duration from the serving cell assistance information.

The validity duration parameter applies to both connected and idle mode UEs. The UE in RRC_IDLE and RRC_INACTIVE shall ensure having a valid version of SIB19 if UE is accessing NR via NTN access. Upon receiving SIB19, the UE in RRC_CONNECTED shall: start or restart T430[14] for

---

13 In Release-17, only first-order Taylor polynomial/linear approximation and second-order Taylor polynomial/ quadratic approximation can be used for Common TA calculation: Indeed, only ta-CommonDrift (1st order derivative) and ta-CommonDriftVariation (2nd order derivative) can be provided. Indicating also Common TA third order derivative was discussed during the normative phase but not agreed.
14 T430 timer associated with the validity duration. It starts or restarts from the subframe indicated by epochTime upon reception of SIB19, or upon reception of RRCReconfiguration message for the target cell including reconfigurationWithSync, or upon conditional reconfiguration execution, i.e., when applying a stored RRCReconfiguration message for the target cell including reconfigurationWithSync.

Table 5.8 Epoch time and validity duration parameters.

Parameter name	Description
Epoch time and validity duration parameters:	
EpochTime	Indicate the epoch time for assistance information by an SFN and a sub-frame number.
	Value range: 0–1023 to indicate SFN and 0–9 to indicate the sub-frame number.
ntn-UlSyncValidityDuration	A validity duration configured by the network for assistance information (i.e. Serving and/or neighbor satellite ephemeris and Common TA parameters), which indicates the maximum time duration (from **epochTime**) during which the UE can apply assistance information without having acquired new assistance information.
	The unit of **ntn-UlSyncValidityDuration** is second.
	Value *s5* corresponds to 5 s, value *s10* indicates 10 s, and so on.

serving cell with the timer value set to **ntn-UlSyncValidityDuration** for the serving cell from the subframe indicated by epoch time for the serving cell. UE should attempt to reacquire SIB19 before the end of the duration indicated by **ntn-UlSyncValidityDuration** and **epochTime** by UE implementation (this is what UE2 did in the example in Figure 5.13). The UE assumes that it has lost uplink in synchronization if new assistance information is not available within the validity duration (This is what happened to UE1 in the example of Figure 5.13). As illustrated in Figure 5.14, the exact time when UL synchronization is obtained (after SIB19 is acquired) is left to UE implementation, which can be from the subframe indicated by epochTime and optionally before the subframe indicated by epochTime. Thereby, the UE may consider assistance information valid as soon as it is received.

5.4.1.4 Timing Advance Adjustment Delay

In TN, as defined in 3GPP TS 38.133 [14], UE shall adjust the timing of its uplink transmission timing at time slot $n + k + 1$ for a TA command received in time slot n, and the value of k is defined in clause 4.2 in TS 38.213 [6]. The same requirement applies also when the UE is not able to transmit a configured uplink transmission due to the channel assessment procedure.

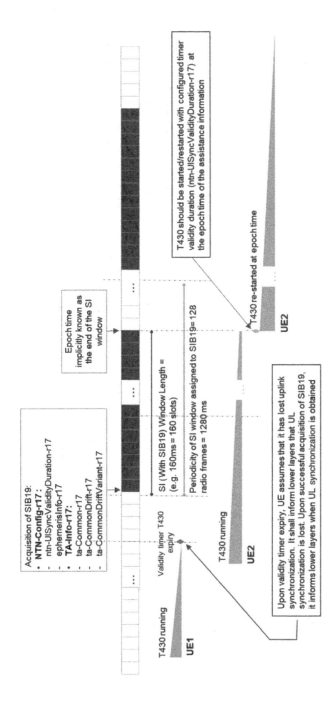

Figure 5.13 Uplink synchronization validity duration and associated UE behavior.

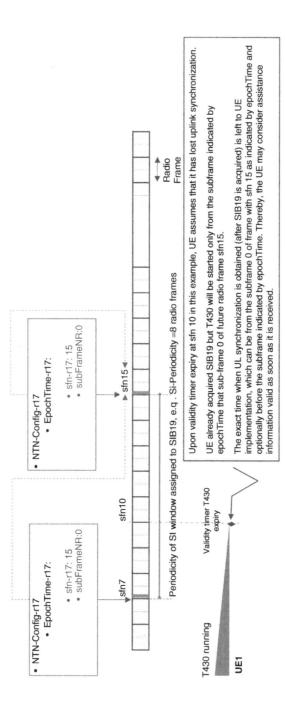

Figure 5.14 Time when UL synchronization is obtained after SIB19 is acquired.

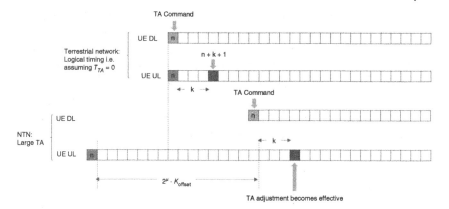

Figure 5.15 Timing Advance adjustment in NR NTN.

In NTN, a UE may need to apply a large TA value that leads to a large offset in its DL and UL frame timing. The existing NR timing definitions involving DL-UL timing interaction may not hold with such large offset in the UE's DL and UL frame timing in NTN. Therefore, 3GPP Release-17 specifications introduce **Koffset** to enhance the adjustment of uplink transmission timing upon the reception of a corresponding TA command. Koffset was introduced as part of NR timing relationships enhancements for NR NTN as illustrated in Figure 5.15 and will be detailed in Section 5.5.

For a TA command received on uplink slot n and for a transmission other than a PUSCH scheduled by a RAR UL grant or a fallbackRAR UL grant, or a PUCCH with HARQ-ACK information in response to a successRAR, the corresponding adjustment of the uplink transmission timing applies from the beginning of uplink slot $n + k + 1 + 2^\mu \cdot K_{offset}$, where $K_{offset} = K_{cell,offset} - K_{UE,offset}$, where $K_{cell,offset}$ is provided by **Koffset** in ServingCellConfigCommon and $K_{UE,offset}$ is provided by a MAC CE command; otherwise, if not respectively provided, $K_{cell,offset} = 0$ or $K_{UE,offset} = 0$.

5.4.2 Uplink Frequency Synchronization

In 3GPP Release-17, NTN UE shall support frequency pre-compensation to counter shift the Doppler experienced on the service link. As part of UE capability parameters, the field **uplinkPreCompensation-r17** is used to indicate whether the UE supports the uplink time and frequency pre-compensation and timing relationship enhancements. As specified in 3GPP TS 38.306 [12], support of uplink frequency pre-compensation in NTN bands is mandatory for UE supporting NR NTN access. The maximum

Doppler shift and Doppler shift variation to be supported in NTN scenarios are given in Table 5.9.[15]

The NTN UE pre-compensates the uplink-modulated carrier frequency by the estimated Doppler shift before transmitting on the UL. Using its GNSS-acquired UE position $\overrightarrow{X_{UE}}$ and the satellite velocity $\overrightarrow{V_{Sat}}$ obtained from the satellite ephemeris indicated by the network, e.g., in SIB19, the UE can derive the radial velocity between the UE and the satellite $\overrightarrow{U_{Sat\_UE}}$. Then, the radial velocity is used to calculate the Doppler (in ppm) experienced on the service links as follows:

$$f_{Doppler,UL} = \frac{\langle \overrightarrow{V_{Sat}}, \overrightarrow{U_{Sat\_UE}} \rangle}{c \| \overrightarrow{U_{Sat\_UE}} \|}$$

where $\langle \overrightarrow{V_{Sat}}, \overrightarrow{U_{Sat\_UE}} \rangle$ is the dot vector product of $\overrightarrow{V_{Sat}}$ and $\overrightarrow{U_{Sat\_UE}}$. $\| \overrightarrow{U_{Sat\_UE}} \|$ is the Euclidean norm of $\overrightarrow{U_{Sat\_UE}}$. And c is the speed of light.

Note that the UE uses the Downlink signals as frequency synchronization source. The GNSS receiver at the UE is only used to obtain the UE position $\overrightarrow{X_{UE}}$ and not used as a synchronization source.

In NR NTN, as specified in [19], the modulated carrier frequency error at the UE side shall be within ± 0.1 PPM[16] observed over a period of 1 ms compared to ideally pre-compensated reference uplink carrier frequency. The ideally pre-compensated reference uplink carrier frequency consists of the UL carrier frequency signaled to the UE by Satellite Access Node

Table 5.9 Max Doppler shift/drift to be supported in NTN scenarios [2].

Scenarios	GEO-based NTN access	LEO-based NTN access
Max Doppler shift (earth fixed UE)	0.93 ppm	24 ppm (600 km) 21 ppm (1200 km)
Max Doppler shift variation (earth fixed UE)	0.000 045 ppm/s	0.27 ppm/s (600 km) 0.13 ppm/s (1200 km)

Source: TR 38.821 [2].

---

15 It is also mentioned in [[1], TR 38.811, section 7.3.2.4.1] that the Doppler drift is -544 Hz/s @ 2 GHz. This implies that in 1 second, the Doppler shift drift is -544 Hz, or in one LEO RTD = 25.77 ms, the Doppler shift drift is -14.02 Hz ($= -544$ Hz * 25.77/1000).

16 The frequency error corresponding to ± 0.1 ppm is ± 200 Hz with carrier frequency of 2 GHz, ± 2 kHz with carrier frequency of 20 GHz and ± 3 kHz with carrier frequency of 30 GHz.

(SAN) and UL pre-compensated Doppler frequency shift. This requirement includes all the frequency errors, including frequency tracking errors, crystal oscillator drift, and the error on radial velocity calculation. 3GPP Release-17 specifies for an NTN UE to pre-compensate the Doppler experienced on the service link before any uplink transmission in RRC_IDLE and RRC_INACTIVE states (before PRACH transmission) and in RRC_CONNECTED state. This can be seen as an open-loop to maintain UL frequency synchronization. It was discussed during the normative phase to introduce a new closed-loop mechanism for frequency adjustment as it is currently specified for uplink timing control. But this was not agreed. Indeed, by considering the maximum Doppler shift variation to be supported in NTN as shown in Table 5.9, the satellite velocity accuracy shown in Table 5.6 (Section Satellite Ephemeris Information), a connected UE can autonomously predict and pre-compensate the Doppler shift drift with sufficient accuracy before transmitting on the Uplink.

5.5 NR Timing Relationships Enhancements for NR NTN

The timing relationships in the 5G NR physical layer involving DL–UL timing interaction such as the timing of the scheduling of PUSCH, PUCCH, and so on were designed mainly for terrestrial mobile communications, in which the propagation delays are typically less than 1 ms. However, in NTN, as discussed in Section 5.4.1, a UE may need to apply a large TA value that leads to a large offset in its DL and UL frame timing. Thereby, the existing NR timing definitions may not hold. As a consequence, various NR physical layer timing relationships, including uplink resource allocation timings, UE processing and preparation procedure timings, and medium access control (MAC) control element (CE) activation timings, are enhanced in 3GPP Release-17 by introducing two additional slot offsets referred to as **Koffset** and **kmac**.

5.5.1 Timing Relationships Enhanced With Koffset

The following timing relationship enhancements have been introduced in 3GPP Release-17 with the support of Koffset:

- The transmission timing of RAR grant or fallbackRAR grant scheduled PUSCH: Refer to Section 5.5.1.1

- The transmission timing of PDCCH ordered PRACH: Refer to Section 5.5.1.1
- The transmission timing of DCI scheduled PUSCH, including CSI transmission on PUSCH: Refer to Section Transmission Timing of PUSCH in NTN (Figure 5.16)
- The timing of the first PUSCH transmission opportunity in type-2 configured grant: Refer to Section Transmission Timing of PUSCH in NTN
- The transmission timing of HARQ-ACK on PUCCH, including HARQ-ACK on PUCCH to message B (MsgB) in two-step random access: Refer to Section 5.6.4
- The timing of the adjustment of uplink transmission timing upon reception of a corresponding TA command: Refer to Section 5.4.1.4
- The transmission timing of aperiodic SRS: refer to Section 5.5.1.3
- The CSI reference resource timing: refer to Section 5.5.1.3

Because the Koffset is used in several timing relationships in initial random access, Release-17 defined a cell-specific Koffset that needs to be indicated in SI. This cell-specific K_{offset} is provided in **ntn-Config** in SIB19 by the parameter **cellSpecificKoffset**. The value range of the cell-specific K_{offset} is 0–1023 ms, which is sufficient to accommodate different satellite orbits (Table 5.10).

Update of Koffset after initial access is also supported. To accommodate for UE-specific propagation delay to further improve scheduling efficiency, a UE-specific Koffset[17] value can be provided and updated by network with a MAC CE after initial access. Thereby, the additional slot offset K_{offset} equals $K_{cell,offset} - K_{UE,offset}$ where $K_{cell,offset}$ is provided by **cellSpecificKoffset** and

Table 5.10 Definition of cell-specific Koffset.

cellSpecificKoffset
Scheduling offset used for the timing relationships that are modified for NTN. The unit of the field Koffset is number of slots for a given subcarrier spacing of 15 kHz. If the field is absent, UE assumes value 0.
cellSpecificKoffset-r17   INTEGER(1..1023)

---

[17] As part of UE capability **ue-specific-K-Offset-r17** parameter indicates whether the UE supports the reception of UE-specific Koffset comprised of the following functional components: Support of reception of UE-specific Koffset via MAC-CE, Support of determining the timing of PUSCH, PUCCH, CSI reference resource, transmission of aperiodic SRS, activation of TA command, first PUSCH transmission in CG Type 2 with UE-specific Koffset.

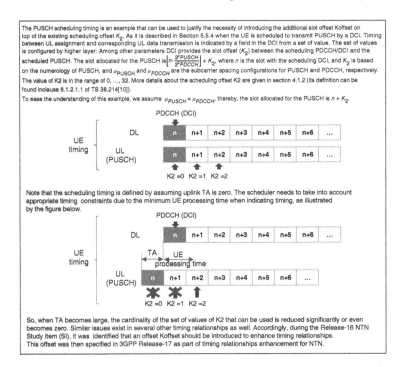

The PUSCH scheduling timing is an example that can be used to justify the necessity of introducing the additional slot offset Koffset on top of the existing scheduling offset K_2. As it is described in Section 5.5.4 when the UE is scheduled to transmit PUSCH by a DCI. Timing between UL assignment and corresponding UL data transmission is indicated by a field in the DCI from a set of value. The set of values is configured by higher layer: Among other parameters DCI provides the slot offset (K_2) between the scheduling PDCCH/DCI and the scheduled PUSCH. The slot allocated for the PUSCH is $\left\lfloor n \cdot \frac{2^{\mu_{PUSCH}}}{2^{\mu_{PDCCH}}} \right\rfloor + K_2$, where n is the slot with the scheduling DCI, and K_2 is based on the numerology of PUSCH, and μ_{PUSCH} and μ_{PDCCH} are the subcarrier spacing configurations for PUSCH and PDCCH, respectively. The value of K2 is in the range of 0, ..., 32. More details about the scheduling offset K2 are given in section 4.1.2 (its definition can be found inclause 6.1.2.1.1 of TS 38.214[10]).

To ease the understanding of this example, we assume $\mu_{PUSCH} = \mu_{PDCCH}$, thereby, the slot allocated for the PUSCH is $n + K_2$.

Note that the scheduling timing is defined by assuming uplink TA is zero. The scheduler needs to take into account appropriate timing constraints due to the minimum UE processing time when indicating timing, as illustrated by the figure below.

So, when TA becomes large, the cardinality of the set of values of K2 that can be used is reduced significantly or even becomes zero. Similar issues exist in several other timing relationships as well. Accordingly, during the Release-16 NTN Study Item (SI), it was identified that an offset Koffset should be introduced to enhance timing relationships. This offset was then specified in 3GPP Release-17 as part of timing relationships enhancement for NTN.

Figure 5.16 An example to justify the necessity of introducing an additional slot offset (K_{offset}).

$K_{UE,offset}$ is provided by a differential Koffset MAC CE command as specified in [TS 38.321]. Both **cellSpecificKoffset** and UE-specific differential Koffset are indicated in the number of slots using SCS of 15 kHz.

The value range of the cell-specific **cellSpecificKoffset** is 0–1023 ms, which is sufficient to accommodate different types of satellite deployment scenarios. The UE-specific differential Koffset is coded in 6 bits, which provides a range from 0–63 ms. The differential Koffset MAC CE is shown in Figure 5.17.

When UE is not provided with UE-specific differential Koffset value, the cell-specific Koffset value, i.e., **cellSpecificKoffset** signaled in SIB19 is used in all the timing relationships that require Koffset enhancements;

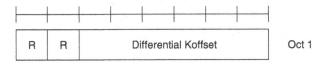

Figure 5.17 Differential Koffset MAC Control Element.

That is, $K_{offset} = K_{cell,offset}$. When UE is provided with UE-specific differential Koffset value, the $K_{offset} = K_{cell,offset} - K_{UE,offset}$ is used in all the timing relationships that require additional slot offset K_{offset} enhancement, except for the following timing relationships, where the cell-specific Koffset value signaled in SIB19 is always used to avoid ambiguity between network and UE about the value of Koffset applied:

- The transmission timing of PUSCH scheduled by RAR UL grant and fallbackRAR grant scheduled PUSCH.
- The transmission timing of message 3 (Msg3) retransmission scheduled by DCI format 0_0 with cyclic redundancy check (CRC) scrambled by temporary cell radio network temporary identifier (TC-RNTI).
- The transmission timing of HARQ-ACK on PUCCH to contention resolution of the PDSCH scheduled by DCI format 1_0 with CRC scrambled by TC-RNTI.
- The transmission timing of HARQ-ACK on PUCCH to MsgB scheduled by DCI format 1_0 with CRC scrambled by MsgB radio network temporary identifier (MsgB-RNTI).
- The transmission timing of PDCCH ordered PRACH.

5.5.1.1 Random-access Procedure in NTN
Random-access procedure is triggered at initial access from RRC_IDLE, but also by a number of other events, for instance, when UL synchronization is lost, during handover execution, for beam failure recovery (BFR) (refer to Section 5.5.2.2) and other triggering events listed in section 9.2.6 of TS 38.300 [4].

The random-access procedure can be contention-based (e.g. at initial connection from idle mode) or non-contention-based (e.g. during Handover to a new cell). Random-access resources and parameters are configured by the network and signaled to the UE (via broadcast or dedicated signaling).

From the physical layer perspective, L1 random-access procedure is triggered upon request of a PRACH transmission by higher layers or by a PDCCH order. The four-step random-access procedure is initiated with the transmission of random-access preamble (Msg1) in a PRACH, followed by a RAR message with a PDCCH/PDSCH (Msg2), and when applicable, the transmission of a PUSCH (Msg3) scheduled by a RAR UL grant, and PDSCH for contention resolution (Msg4). The two-step random-access procedure includes the transmission of random-access preamble in a PRACH and of a PUSCH (MsgA) and the reception of a RAR message with a PDCCH/PDSCH (MsgB), and when applicable, the transmission of a PUSCH scheduled by a fallback RAR UL grant, and PDSCH for contention resolution. The four-step and two-step RACH procedures are shown in Figure 5.18.

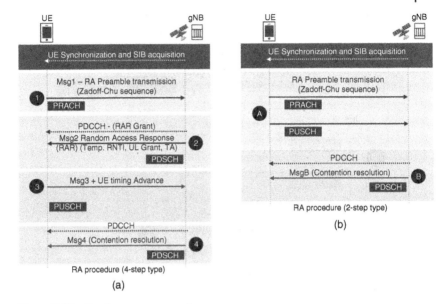

Figure 5.18 Random access procedure.

Fallback for random access with two-step RA type is also supported: If fallback indication is received in MsgB, the UE performs Msg3 transmission using the UL grant scheduled in the fallback indication and monitors contention resolution as shown in Figure 5.19.

Prior to 3GPP Release-17, for the transmission of PUSCH scheduled by the RAR UL grant received in mgs2 (Figure 5.18) or by the fallbackRAR received in MsgB (Figure 5.19) if the UE receives a PDSCH with a RAR (or fallback-RAR) message ending in slot n for a corresponding PRACH transmission from the UE, the UE transmits the PUSCH in slot $n + k_2 + \Delta$. The slot offset k_2 is defined in Section 5.5.1 (refer to Figure 5.16). Δ is an additional slot delay value specific to the PUSCH SCS μ_{PUSCH}, which is applied in addition to the K_2 value for the first transmission of PUSCH scheduled by the RAR or by the fallbackRAR. The value of Δ is defined in clause 6.1.2.1.1 of [[20], TS 38.214] and it equals two slots in case of 15 kHz and three slots in case of 30 kHz SCS of the PUSCH.

For NTN-based access in Release-17, for the transmission of PUSCH scheduled by the RAR UL grant or by the fallbackRAR the **cellSpecificK-offset** is added as an additional slot delay value on top of the existing slot offsets k_2 and Δ. Thereby, if a UE receives a PDSCH with a RAR message ending in slot n for a corresponding PRACH transmission from the UE, the UE transmits the PUSCH in slot $n + k_2 + \Delta + 2^\mu \cdot K_{cell,offset}$, where $K_{cell,offset}$ is provided by **cellSpecificKoffset** in **ntn-Config** carried in SIB19.

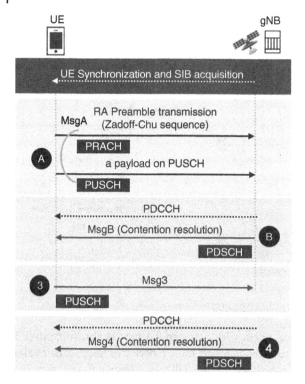

Figure 5.19 Fallback for RA with 2-step RA type.

The random-access procedure can be triggered by a PDCCH order at DL or UL data arrival for a UE, which has lost uplink synchronization while in RRC_CONNECTED or during RRC_INACTIVE while Small Data Transmission (SDT) procedure is ongoing. The PDCCH order triggers the UE to initiate the RA procedure to re-synchronize the UE before forwarding the data to the UE. For a PRACH transmission by a UE triggered by a PDCCH order, the PRACH mask index field [11], if the value of the random-access preamble index field is not zero, indicates the PRACH occasion for the PRACH transmission where the PRACH occasions are associated with the SSB index indicated by the SSB index field of the PDCCH order. If the UE is provided $K_{cell,offset}$ by **cellSpecificKoffset** in **ntn-Config** carried in SIB19, the PRACH occasion is after slot $n + 2^{\mu} \cdot K_{cell,offset}$ where n is the slot of the UL BWP for the PRACH transmission that overlaps with the end of the PDCCH order reception assuming $T_{TA} = 0$, and μ is the SCS configuration for the PRACH transmission.

After Msg1–RA Preamble transmission, the UE monitors the PDCCH for the RAR message (Msg2). The time window to monitor RA response is controlled by higher layers parameter **ra-ResponseWindow,** which starts at a determined time interval after the preamble transmission as specified in section 8.2 of TS 38.213 [6]. If no valid response is received during the **ra-ResponseWindow,** a new preamble is sent. If more than a certain number of preambles have been sent, a random-access problem will be indicated to upper layers. Similarly, the time window to monitor RA response for two-step RA type is controlled by **msgB-ResponseWindow,** which starts at a determined time interval after msgA transmission as specified in section 8.2A of TS 38.213 [6].

In TN, the RAR is expected to be received by the UE within a few milliseconds after the transmission of the corresponding preamble. Thereby, the network configures a value of **ra-ResponseWindow** lower than or equals 10 ms when Msg2 is transmitted in licensed spectrum and a value lower than or equals 40 ms when Msg2 is transmitted with shared spectrum channel access (see TS 38.321, clause 5.1.4). For **msgB-ResponseWindow,** the network configures a value lower than or equals 40 ms (see TS 38.321, clause 5.1.1).

In NTN, the propagation delay is much larger, and thereby, the RAR message cannot be received by the UE within the specified time interval specified for terrestrial communications. Therefore, the starts of **ra-ResponseWindow** and **msgB-ResponseWindow** are delayed by an additional offset corresponding to UE-gNB RTT as illustrated in Figure 5.20. The estimate of UE-gNB RTT equals $T_{TA} + k_{mac}$ ms. where T_{TA} is the TA applied by the UE. And k_{mac} is provided by **kmac** within **ntn-Config** in

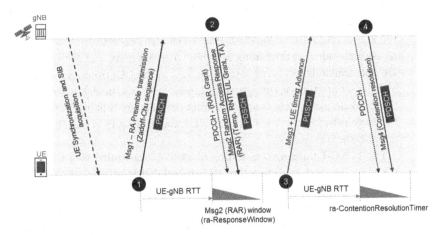

Figure 5.20 Msg2 (RAR) window and ra-ContentionResolution delayed by UE-gNB RTT.

SIB19 or $k_{mac} = 0$ if **kmac** is not provided. The value of k_{mac} equals the offset of the gNB's downlink and uplink frame timing as will be discussed in Section 5.5.2. Note that this additional delay introduced for the starts of **ra-ResponseWindow** and **msgB-ResponseWindow** should only apply to NTN.

When the UE sends an RRC Connection Request (Msg3), it will monitor for Msg4 in order to resolve a possible random-access contention. The **ra-ContentionResolutionTimer** starts after Msg3 transmission. The maximum configurable value of **ra-ContentionResolutionTimer** is large enough to cover the RTD in NTN. However, to save UE power, the behavior of the contention resolution timer is also modified to support NTN: It is also delayed by UE-gNB RTT.

5.5.1.2 Resource Allocation in Time Domain

Resource Allocation in Time Domain in TN In 5G NR, the PDCCH is used to schedule DL transmissions on PDSCH and UL transmissions on PUSCH. The DCI on PDCCH includes:

- Downlink assignments containing at least modulation and coding format, resource allocation, and hybrid-ARQ information related to DL-SCH
- Uplink scheduling grants containing at least modulation and coding format, resource allocation, and hybrid-ARQ information related to UL-SCH

For PDSCH recourse allocation, when the UE is scheduled to receive PDSCH by a DCI, the DCI indicates the slot offset K_0. The slot allocated for the PDSCH is $\left\lfloor n \cdot \frac{2^{\mu_{PDSCH}}}{2^{\mu_{PDCCH}}} \right\rfloor + K_0$, where n is the slot with the scheduling DCI, K_0 is based on the numerology of PDSCH, and μ_{PDSCH} and μ_{PDCCH} are the SCS configurations for PDSCH and PDCCH, respectively. The value of K_0 corresponding to the timing between DL assignment and corresponding DL data transmission is indicated by a field in the DCI from a set of values and the set of values is configured by higher layer. As the PDSCH reception timing is defined solely from DL timing perspective. It is not impacted by the large offset in the UE's DL and UL frame timing in NTN. Thereby no enhancement is needed.

The PUSCH transmission timing of PUSCH scheduled by DCI is illustrated in Figure 5.21. When the UE is scheduled to transmit PUSCH by a DCI, the DCI indicates the slot offset K_2. The slot allocated for the PUSCH is $\left\lfloor n \cdot \frac{2^{\mu_{PUSCH}}}{2^{\mu_{PDCCH}}} \right\rfloor + K_2$, where n is the slot with the scheduling DCI, K_2 is based on the numerology of PUSCH, and μ_{PUSCH} and μ_{PDCCH} are the SCS configurations for PUSCH and PDCCH, respectively. The value of K_2 is in the range of 0, ..., 32. PUSCH may be scheduled with DCI on PDCCH, or a

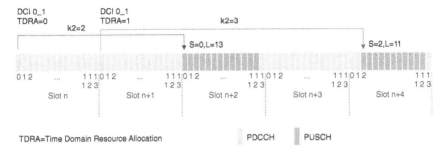

Figure 5.21 Transmission timing of PUSCH in TN.

semi-statically configured grant may be provided over RRC. By considering the large TA in NTN, and the limited range of K_2, the PUSCH transmission timing of PUSCH needs to be enhanced in NTN as will be discussed in Section Transmission Timing of PUSCH in NTN.

Transmission Timing of PUSCH in NTN In NTN, The UE needs to apply a large TA value to compensate for the RTT between UE and the UTSRP (refer to Section 5.4.1.2). By considering this large TA in NTN and as the maximum value for K_2 equals 32, it is clear that the existing range K_2 is not enough for PUSCH resource allocation in NTN. Indeed, even by using the maximum value of K_2 of 32, from the UE's perspective, the uplink slot $n + K_2$, where PUSCH is supposed to be transmitted may occur before the downlink slot n where the scheduling DCI is received. To resolve this issue, the scheduling offset, i.e., Koffset (refer to Section 5.5.1) is used to enhance the PUSCH transmission timing. Specifically, for a scheduling DCI received in downlink slot n, the uplink slot where the UE shall transmit the PUSCH is determined as slot $m = \left\lfloor n \cdot \frac{2^{\mu_{PUSCH}}}{2^{\mu_{PDCCH}}} \right\rfloor + K_2 + K_{offset} \cdot \frac{2^{\mu_{PUSCH}}}{2^{\mu_{K_{offset}}}}$. With appropriate value of K_{offset}, the uplink slot m can occur after the downlink slot n at the UE side. Where, as specified in TS 38.213 [6], K_{offset} is configured by higher-layer **cellSpecificKoffset** and may be updated by a differential Koffset provided by MAC CE command, and where $\mu_{K_{offset}}$ is the SCS configuration for K_{offset} with a value of 0 for FR1, n is the slot with the scheduling DCI, K_2 is based on the numerology of PUSCH, μ_{PUSCH} and μ_{PDCCH} are the SCS configurations for PUSCH and PDCCH, respectively.

In FR1, if the same SCS configurations for PUSCH and PDCCH are used, the slot where the UE shall transmit the PUSCH is determined by $m = n + K_2 + K_{offset} \cdot 2^{\mu_{PUSCH}}$. In case of PUSCH transmission with $SCS = 15\,\text{kHz}$ as illustrated in Figure 5.22, $m = n + K_2 + K_{offset}$. When $SCS = 30\,\text{kHz}$, $m = n + K_2 + K_{offset} \cdot 2$.

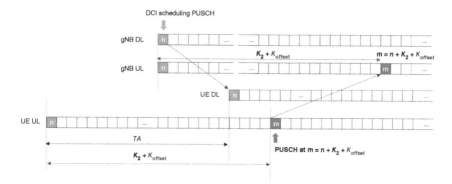

Figure 5.22 Transmission timing of DCI scheduled PUSCH in NTN using Koffset in FR1 and SCS = 15 khz.

In 5G NR, with UL Configured Grants, the scheduler can allocate uplink resources to the UE. When a configured uplink grant is active, if the UE cannot find an uplink grant assigned via downlink control channel an uplink transmission according to the configured uplink grant can be made. Otherwise, if the UE finds an uplink grant assigned via downlink control channel, this assignment overrides the configured uplink grant. Two types of configured grants supported for UL in NR, Type 1 and Type 2. With Type 1, the configured uplink grant-related parameters (including periodicity, PRBs, and MCS) are provided via RRC without DCI activation/deactivation. In this type of operation, the PUSCH is triggered by data arrival to the UE's transmit buffer and the PUSCH transmissions follow the RRC configuration. With Type 2, the periodicity of the configured uplink grant periodicity is defined via RRC, but PRBs, MCS, and so on are activated and deactivated via DCI. In this type of operation, the first PUSCH is triggered with a DCI, with subsequent PUSCH transmissions following the RRC configuration and scheduling received on the DCI.

In 5G NTN, as specified in TS 38.214 [20], Koffset can be applied to the first transmission opportunity of PUSCH in Configured Grants Type 2 in the same way as Koffset is applied to the transmission timing of DCI scheduled PUSCH described above. Further, Koffset is not introduced for Type 1 configured grant in 3GPP Release-17 NTN.

5.5.1.3 Other Timing Relationships Enhanced With Koffset

The Transmission Timing of Aperiodic SRS SRS are uplink physical signals that can be used for estimation of uplink CSI to assist uplink scheduling, uplink power control, as well as assist the downlink transmission (e.g., the downlink beamforming in the scenario with UL/DL reciprocity). The resource allocation of SRS in the time domain and SRS frequency allocation are specified in clause 6.4.1.4 of TS 38.211 [5]. The SRS is configured using one or more **SRS-ResourceSets** and **SRS-PosResourceSets**, where each resource set defines a set of **SRS-Resources** or **SRS-PosResources**. The IE **SRS-Config** which is used to configure SRS transmissions is described in TS 38.331 [15]. SRS transmission timing is semi-statically configurable by higher-layer parameter **resourceType**, which may be periodic, semi-persistent, or aperiodic SRS transmission:

- An aperiodic SRS transmission is triggered using DCI on the PDCCH.
- Semi-persistent SRS activation is triggered by MAC CE; once triggered, it becomes a periodic transmission until a MAC CE is received to stop it.
- A periodic SRS is transmitted with a certain configured periodicity (in number of slots) and a certain configured slot offset within that periodicity. Periodic SRS transmission does not require an activation instruction after the UE receives the SRS Resource Set configuration.

For NTN-based access, with regard to SRS transmission, only aperiodic SRS transmission timing needs to be enhanced with Koffset: For aperiodic SRS transmission, a slot level offset is defined by the higher-layer parameter **slotOffset**, which is an offset in number of slots between the triggering DCI and the actual SRS transmission. The UE applies no offset if the field is absent. As specified in clause 6.2.1 of 3GPP TS 38.214 [20], if a UE receives a DCI triggering aperiodic SRS in slot n, the UE transmits aperiodic SRS in each of the triggered SRS resource set(s) in slot $K_s = \left\lfloor n \cdot \frac{2^{\mu_{SRS}}}{2^{\mu_{PDCCH}}} \right\rfloor + k +$ $K_{offset} \cdot \frac{2^{\mu_{SRS}}}{2^{\mu_{K_{offset}}}}$, where k is configured using the parameter **slotOffset** for each triggered SRS resources set and is based on the SCS of the triggered SRS transmission, μ_{SRS}, and μ_{PDCCH} are the SCS configurations for triggered SRS and PDCCH carrying the triggering command, respectively. K_{offset}[18] is the parameter configured by higher layer as described in Section 5.5.1. $\mu_{K_{offset}}$ is the SCS configuration for Koffset with a value of 0 for FR1.

In FR1 and if the triggered SRS and the PDCCH carrying the triggering are configured with the same SCS, the above formula can be simplified as: $K_s = n + k + K_{offset}$.

The CSI Reference Resource Timing CSI-RS is used in 5G NR for the estimation of CSI in downlink to further prepare feedback reporting to the Base Station to assist in MCS selection, beamforming, MIMO rank selection, and

---

18 As described in Section 5.5.1, K_{offset} equals $K_{cell,offset} - K_{UE,offset}$, where $K_{cell,offset}$ is provided by **cellSpecificKoffset** and $K_{UE,offset}$ is provided by a differential Koffset MAC CE command.

resource allocation. CSI-RS transmissions are transmitted periodically, in an aperiodic way, and semi-persistently at a configurable rate by the Base Station. CSI-RS also can be used for interference measurement and fine frequency/time tracking purposes.

The CSI reference resource is a group of downlink frequency-domain and time-domain resources that are configured as defined in TS 38.214 [20], section 5.2.2.5. The CSI reference resource timing is another timing relationship involving DL-UL timing interaction that needs to be enhanced for NTN. According to 3GPP TS 38.214 [20], the CSI reference resource for a CSI report in uplink slot n' is defined by a single downlink slot $n - n_{CSI\_ref} - K_{offset} \cdot \frac{2^{\mu_{DL}}}{2^{\mu_{K_{offset}}}}$, where $n = \left\lfloor n' \cdot \frac{2^{\mu_{DL}}}{2^{\mu_{UL}}} \right\rfloor$, μ_{DL} and μ_{UL} are the SCS configurations for DL and UL, respectively. The value of n_{CSI_ref} depends on the type of CSI report and is defined in TS 38.214 [20]. K_{offset} is the parameter configured by higher layer as described in Section 5.5.1.

5.5.2 Timing Relationships Enhanced With kmac

The main services and functions of the MAC sublayer include, among others as specified in TS 38.321, multiplexing of MAC SDUs from one or different logical channels onto TB to be delivered to the physical layer on transport channels. And de-multiplexing of MAC SDUs to one or different logical channels from TB delivered from the physical layer on transport channels. Further, the MAC layer can also insert MAC CE into the TBs to be transmitted over the transport channels. MAC CEs are used for in-band control signaling and provide a faster way to send control information than RLC. MAC PDU structure is described in Figure 5.23: as shown in this figure, MAC PDU includes the MAC subPDU with fixed size and variable size MAC CE.

There are multiple MAC CEs in NR used for various purposes such TAC sent by gNB to adjust UE timing, differential Koffset MAC CE as discussed in Section 5.5.1. And many other MAC CEs, which can be found in 3GPP TS 38.321. Some of these MAC CEs (e.g. the Recommended Bit Rate MAC CE) do not involve timing relationships defined in the physical layer specifications.

In essence, for MAC CE timing relationships defined in the physical layer specification, the MAC CE command becomes activated 3 ms after UE transmits HARQ-ACK corresponding to the received PDSCH carrying the MAC CE command. In fact, when the UE transmits a PUCCH with HARQ-ACK information in uplink slot n corresponding to a PDSCH carrying a MAC CE command, the UE action and assumption on downlink or uplink configuration indicated by the MAC CE command starts from the first downlink slot after downlink slot $n + 3N_{slot}^{subframe,\mu}$; μ is the SCS configuration

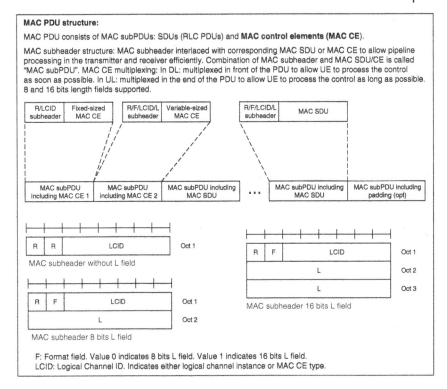

Figure 5.23 MAC PDU structure.

for the PUCCH and $N_{slot}^{subframe,\mu}$ is the number of slots per subframe for SCS configuration μ. This means the MAC CE action has an application delay of 3 ms. Note that the MAC CE action timing in LTE standard has also an application delay of 3 ms. It is worth noting also that there are exceptions to the rule discussed above, such as the adjustment of the uplink transmission timing corresponding to a TA MAC CE command as discussed in Section 5.4.1.4.

To ease the understanding of the MAC CE timing relationships enhanced with **kmac**, let's consider the following examples: Figure 5.24 illustrates MAC CE timing relationships in TN. Figures 5.25 and 5.26 illustrate respectively the MAC CE timing relationships in NTN-based access when DL and UL frame timing are aligned at gNB and when DL and UL frame timing are not aligned at gNB. In these three examples, slot n denotes the slot at which the PUCCH with HARQ-ACK acknowledging the PDSCH carrying MAC CE Command is received by the Base Station. And the slot m is the slot where the UL or DL MAC CE command is activated, which is determined by the MAC CE action application delay of $3N_{slot}^{subframe,\mu}$ (or 3 ms).

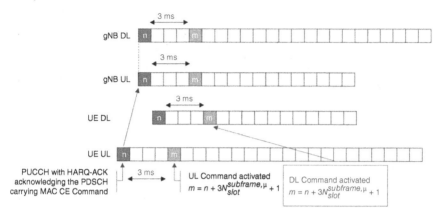

Figure 5.24 MAC CE timing relationships in TN.

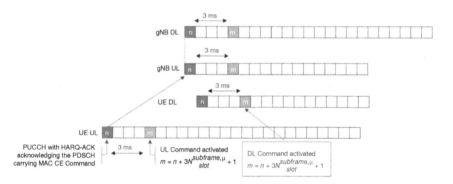

Figure 5.25 MAC CE timing relationships in NTN when DL and UL frame timing are aligned at gNB.

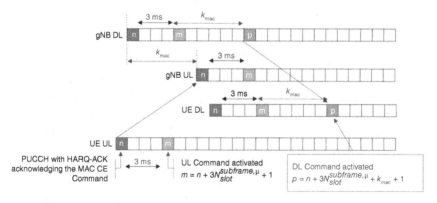

Figure 5.26 MAC CE timing relationship enhancement with K_{mac}.

As illustrated in Figures 5.24 and 5.26, when downlink and uplink frame timing are aligned at the Base Station, slot n arrives always before slot m. Thereby, when the PUCCH with HARQ-ACK acknowledging the PDSCH carrying MAC CE Command is received by the Base Station at slot n, both Base Station and UE have the same understanding that MAC CE command is activated 3 ms later, i.e., in slot m. Note that the main difference between NTN and TN is the large TA value that is applied in NTN. But this does not affect the order of occurrence of slots m and n. Therefore, it was concluded in 3GPP Release-17 that when downlink and uplink frame timing are aligned at the Base Station, the MAC CE timing relationships do not require enhancement.

The example in Figure 5.26 illustrates MAC CE timing relationships when DL and UL frame timing are not aligned at the Base Station. In NTN, this could be the case, for example, when the UTSRP is located onboard the satellite. In this scenario, from the Base Station perspective, the downlink slot $m = n + 3N_{slot}^{subframe,\mu}$ occurs before the actual uplink slot n where the PUCCH with HARQ-ACK information acknowledging the MAC CE command would be received. From the Base Station perspective, as long the downlink MAC CE command is not acknowledged, the UE action and assumption on downlink configuration is not clear and thereby, the slot m cannot anymore be considered as the slot where the DL MAC CE command would be activated.

3GPP Release-17 resolved this issue by enhancing the MAC CE timing relationship with a slot offset referred to as **kmac** in the specifications. This offset is signaled to the UE if downlink and uplink frame timing are not aligned at the Base Station: It is needed for UE action and assumption on downlink configuration indicated by a MAC CE command in PDSCH. If a UE is provided with a **kmac** value, when the UE would transmit a PUCCH with HARQ-ACK information in uplink slot n corresponding to a PDSCH carrying a MAC CE command on a downlink configuration, the UE action and assumption on the downlink configuration shall be applied starting from the first slot that is after slot $n + 3N_{slot}^{subframe,\mu} + 2^{\mu} \cdot k_{mac}$, where μ is the SCS configuration for the PUCCH. As illustrated in Figure 5.26, the value of **kmac** equals the offset between the downlink and the uplink frame timing at the Base Station. The downlink slot $p = n + 3N_{slot}^{subframe,\mu} + 2^{\mu} \cdot k_{mac}$ occurs after the actual uplink slot n where the PUCCH with HARQ-ACK information acknowledging the MAC CE command would be received by Base Station.

Note that for UE action and assumption on uplink configuration indicated by a MAC-CE command in PDSCH, this offset is not needed even if downlink and uplink frame timing are not aligned at the Base Station. This is

Table 5.11 Definition of **kmac**.

kmac

A slot offset is provided by network if downlink and uplink frame timing are not
aligned at gNB. It is needed for UE action and assumption on downlink
configuration indicated by a MAC CE command in PDSCH. If the field is absent
UE assumes value 0.
For the reference subcarrier spacing value for the unit of **kmac** in FR1, a value of
15 kHz is used. The unit of **kmac** is number of slots for a given subcarrier
spacing.

kmac-r17   INTEGER(1..512)

because the uplink slot $m = n + 3N_{slot}^{subframe,\mu}$ always occurs after the actual
uplink slot n where the PUCCH with HARQ-ACK information acknowledg-
ing the MAC CE command would be received.

Obviously, the network does not need to provide **kmac** if downlink and
uplink frame timing are aligned at the Base Station.

kmac is indicated within **ntn-Config** which is broadcast in SIB19 or
provided to the UE in dedicated signaling within **ServingCellConfigCommon**. 3GPP Release-17 specified one value range of **kmac** covering all NTN
deployment scenarios which are 1–512 ms. When UE is not provided by
network with a **kmac** value, UE assumes $k_{mac} = 0$ (Table 5.11).

5.5.2.1 Uplink Power Control

The uplink power control is used in 5G NR to limit both intra-cell
and inter-cell interference and to reduce UE power consumption. Both
open-loop and closed-loop power controls are supported in the uplink.
The uplink power control is independent of uplink data (PUSCH), uplink
control (PUCCH), SRS, and PRACH transmissions as specified in clause 7
of TS 38.213 [6]. The uplink power control is based on both signal-strength
measurements done by the UE itself (open-loop power control), as well as
measurements performed by the Base Station. The latter measurements
are used to generate Transmit power-control (TPC) commands that are
subsequently fedback to the UE as part of the DCI (closed-loop power control). Both absolute and relative power-control commands are supported.
There are four available relative power adjustments ("step size") in case of
relative power control. For uplink data, multiple closed-loop power control
processes can be configured, including the possibility of separate processes
with transmission beam indication. There can be at most four parallel
path-loss estimation processes.

The network configures the UE with a set of downlink reference signals
(CSI-RS or SS block) on which path loss is to be estimated. These RS
resources may be updated by MAC CE (e.g., PUSCH Pathloss Reference

RS Update MAC CE). If one of the RS resources maintained by the UE for pathloss estimation for PUSCH/PUCCH/SRS transmissions is updated by MAC CE Command, the UE applies the pathloss estimation based on the RS resources starting from the first slot that is after slot $k + 3 \cdot N_{\text{slot}}^{\text{subframe, } \mu} + 2^{\mu} \cdot k_{mac}$, where k is the slot where the UE would transmit a PUCCH or PUSCH with HARQ-ACK information for the PDSCH providing the MAC CE, μ is the SCS configuration for the PUCCH or PUSCH, respectively, that is determined in the slot when the MAC CE command is applied and k_{mac} is provided by **kmac** as described in Section 5.5.2 or $k_{mac} = 0$ if **kmac** is not provided.

More details on uplink power control can be found in 3GPP TS 38.213 [6].

5.5.2.2 Beam Failure Recovery

Beam failure recovery (BFR) procedure is specified in clause 6 of TS 38.213 [6]. BFR is used for rapid link reconfiguration against sudden blockages, i.e., fast re-aligning of the Base Station and UE beams. It is a function performed on Physical and MAC protocol layers. MAC initiates BFR procedure when the signal level of serving beams will become very bad (i.e. UE's serving beam fails). Contention-free Random Access (refer to Section 5.5.2) preamble may be allocated by the Network for BFR purpose. If no candidate beam exists for which contention-free preambles have been allocated, UE performs a contention-based random access. BFR procedure is illustrated in Figure 5.27.

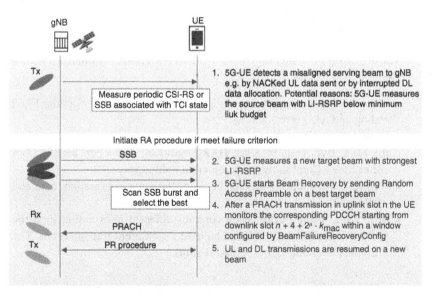

Figure 5.27 The timing relationship for beam failure recovery enhanced with k_{mac}.

In 3GPP Release-17, the **kmac** is used in the BFR, where after a PRACH transmission in uplink slot n the UE monitors the corresponding PDCCH starting from downlink slot $n + 4 + 2^\mu \cdot k_{mac}$ within a window configured by **BeamFailureRecoveryConfig**.[19] Where μ is the SCS configuration for the PRACH transmission.

5.6 Hybrid ARQ Enhancements for NR NTN

In NR, the HARQ mechanism ensures delivery between peer entities at Layer 1. In essence, it provides fast retransmissions on Physical layer, and it is controlled by MAC. On the MAC layer, hybrid ARQ with soft-combining between transmissions is supported. Different redundancy versions (RVs) can be used for different transmissions. The modulation and coding scheme (MCS) may be changed for retransmissions. In order to minimize delay and feedback, a set of parallel stop-and-wait processes are used. To correct possible residual errors, the MAC ARQ is complemented by a robust selective-repeat ARQ protocol on the RLC layer. RLC retransmissions are limited to logical channels in Acknowledged Mode (AM). HARQ offers retransmission capability for RLC Unacknowledged Mode (UM) and Transparent Mode (TM).

More details on the HARQ operation could be found in [4] and [21].

5.6.1 HARQ Functionality Basics

The principles of HARQ functionality in 5G NR TN are summarized within the following bullets:

- The HARQ is a combination of high-rate forward error-correcting (FEC) coding and ARQ error-control.
- Hybrid ARQ with soft-combining between transmissions is supported on the MAC layer (Figure 5.28). HARQ retransmissions can benefit from either Chase Combining or Incremental redundancy (IR).
- With CC, every re-transmission contains the same information, data and parity bits)
- With IR, every re-transmission contains different information than the previous one. Different RVs can be used for different transmissions. The MCS may be changed for retransmissions.

---

19 The IE **BeamFailureRecoveryConfig** is used to configure the UE with RACH resources and candidate beams for beam failure recovery in case of beam failure detection.

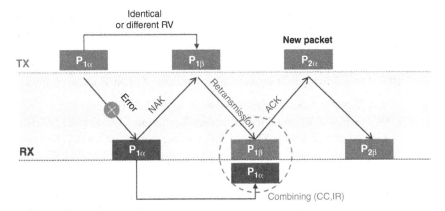

Figure 5.28 Hybrid ARQ principles in NR.

- In order to minimize delay and feedback, a set of parallel stop-and-wait processes are used. Multiple processes permit to compensate for the RTT.
- To correct possible residual errors, the MAC ARQ is complemented by a robust selective-repeat ARQ protocol on the RLC layer.
- Prior to 3GPP Release-17, for downlink, a maximum of **16 HARQ** processes per cell is supported by the UE. The number of processes the UE may assume will at most be used for the downlink is configured to the UE for each cell separately by higher-layer parameter **nrofHARQ-processesForPDSCH**, value from the set {2,4,6,8,10,12,16} [20]. When no configuration is provided, the UE may assume a default number of eight processes.
- Prior to 3GPP Release 17, for uplink, a maximum of **16 HARQ** processes per cell is supported by the UE.
- Each process runs in parallel with the others: It is responsible for generating ACK or NACK indicating delivery status of PDSCH/PUSCH. And it handles STOP and WAIT HARQ protocol. With anticipating window = 1, that is, each process is blocked until the ACK/NACK is received.
- A single HARQ process supports one TB when the physical layer is not configured for downlink/uplink spatial multiplexing.
- A single HARQ process supports one or multiple TBs when the physical layer is configured for downlink/uplink spatial multiplexing.

Dynamic downlink resource allocation is provided via the PDCCH using DCI. The following DCI formats for scheduling of PDSCH are supported:

- DCI format 1_0 is used for the scheduling of PDSCH in one DL cell.
- DCI format 1_1 is used for the scheduling of one or multiple PDSCH in one cell.
- DCI format 1_2 is used for the scheduling of PDSCH in one cell.

The information included within these DCI formats is specified in clause 7.3.1.2 of TS 38.212 [11]. For the operation of DL HARQ, these DCI formats include the key fields described in Table 5.12:

Dynamic uplink resource allocation is provided via the PDCCH using DCI. The following DCI formats for scheduling of PUSCH are supported:

- DCI format 0_0 is used for the scheduling of PUSCH in one cell.
- DCI format 0_1 is used for the scheduling of one or multiple PUSCH in one cell or indicating CG downlink feedback information (CG-DFI) to a UE.
- DCI format 0_2 is used for the scheduling of PUSCH in one cell.

The information included within these DCI formats is specified in clause 7.3.1.1 of TS 38.212 [11]. For the operation of UL HARQ, these DCI formats include the key fields described in Table 5.13:

5.6.2 Increasing the Number of HARQ Processes in NTN

Due to large RTD in NTN, the number of HARQ processes may expire even before the first feedback is received as shown in Figure 5.29. Therefore, to avoid reduction in peak data rates in NTN, increasing the number of HARQ processes avoids HARQ stalling, where a DL packet is received but the corresponding UL HARQ feedback is not yet received and processed in the Base

Table 5.12 Fields related to HARQ in DCI formats for scheduling of PDSCH.

	DCI Format 1_0	DCI Format 1_1	DCI Format 1_2
New Data Indicator (NDI)	1 bit	1 bit	1 bit
Redundancy Version (RV)	2 bits	2 bits	0–2 bits
HARQ process number	4 bits	4 bits	0–4 bits
Downlink Assignment Index (DAI)	2 bits	0, 2, 4, 6 bits	0/1/2/4 bits
PDSCH-to-HARQ feedback timing indicator	3 bits	0–3 bits	0–3 bits
Code Block Group Transmission Information (CBGTI)		0/2/4/6/8 bits	
Code Block Group Flush Indicator (CBGFI)		0/1 bit	
One-shot HARQ-ACK request		0/1 bit	
PDSCH group index		0/1 bit	
New Feedback Indicator (NFI)		0–2 bits	
Number of requested PDSCH groups		0/1 bit	

Table 5.13 Fields related to HARQ in DCI formats for scheduling of PUSCH.

	DCI Format 0_0	DCI Format 0_1	DCI Format 0_2
New Data Indicator (NDI)	1 bit	1 bit	1–8 bits
Redundancy Version (RV)	2 bits	0–2 bits	2–8 bits
HARQ process number	4 bits	0–4 bits	4 bits
Downlink Assignment Index (DAI)		0/1/2/4 bits	0/1/2/4 bits
Code Block Group Transmission Information (CBGTI)		0/2/4/6/8 bits	0/2/4/6/8 bits

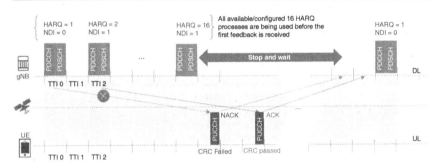

Figure 5.29 HARQ stalling.

Station [13]. This would degrade the performances, particularly the peak throughput. For example, in case of LEO 600 km orbit, with maximum RTT up to 25.77 ms, if only 16 HARQ processes are used, the maximum peak throughput due to HARQ stalling is reduced by about 43%. In such a deployment scenario, increasing the number of HARQ processes is beneficial: 28 HARQ processes are sufficient to avoid HARQ stalling [13].

For NR NTN operation in 3GPP Release-17, the maximal supported HARQ process number is increased, and it is up to 32 for both UL and DL. Specifically, for downlink, a maximum of 16 HARQ processes per cell is supported by the UE, or, subject to UE capability, a maximum of 32 HARQ processes per cell. The number of HARQ processes to be used on the PDSCH of a serving cell is indicated to the UE within **PDSCH-ServingCellConfig**[20] by higher layer parameter **nrofHARQ-ProcessesForPDSCH** if 16 processes or less are to be configured or **nrofHARQ-ProcessesForPDSCH-v1700** if the number of HARQ processes is 32, and when no configuration is provided the UE assumes a default number of 8 processes.

For uplink, 16 HARQ processes per cell are supported by the UE, or, subject to UE capability, a maximum of 32 HARQ processes per cell.

---

20 The IE *PDSCH-ServingCellConfig* is used to configure UE-specific PDSCH parameters that are common across the UE's BWPs of one serving cell.

The number of HARQ processes to be used on the PUSCH of a serving cell is configured to the UE by higher layer parameter **nrofHARQ-ProcessesforPUSCH-r17** within **PUSCH-ServingCellConfig**.[21] With the value n32 corresponds to 32 HARQ processes. The UE uses 16 HARQ processes if the field is absent.

The new and updated higher layer parameters related to number of HARQ processes in 3GPP Release-17 are listed in Table 5.14.

Extension of maximal HARQ process number in Release-17 requires an enhancement on the HARQ process indication: The HARQ process ID field within DCI 0-1/1-1 and DCI 0-2/1-2 is extended to be up to 5 bits when the maximum supported HARQ processes number is configured as 32.

Table 5.14 Number of HARQ processes in NTN.

Parameter name	Parameter description	Value range	Comment
nrofHARQ-ProcessesForPDSCH-r17	The number of HARQ processes to be used on the PDSCH of a serving cell. Value n2 corresponds to 2 HARQ processes, value n4 to 4 HARQ processes, and so on. If the field is absent, the UE uses 8 HARQ processes (see TS 38.214 [20], clause 5.1).	{n2, n4, n6, n10, n12, n16, n32}	Extension of R16
nrofHARQ-ProcessesForPUSCH-r17	The number of HARQ processes to be used on the PUSCH of a serving cell. Value n16 corresponds to 16 HARQ processes, value n32 to 32 HARQ processes. If the field is absent, the UE uses 16 UL HARQ processes (see TS 38.214 [20], clause 5.1).	{n16, n32}	New in Rel-17
nrofHARQ-Processes-r17	The number of HARQ processes configured for UL configured grant. It applies to both Type 1 and Type 2. See TS 38.321, clause 5.4.1.	1...32	Extension of R16
nrofHARQ-Processes-r17	Number of configured HARQ processes for SPS DL. See TS 38.321, clause 5.8.1.	1...32	Extension of R16

---

21 The IE *PUSCH-ServingCellConfig* is used to configure UE-specific PUSCH parameters that are common across the UE's BWPs of one serving cell.

Table 5.15 Enhancement on the HARQ process indication.

Parameter name	Parameter description	Value range	Comment
harq-ProcessNumber SizeDCI-1-2-r17	Configure the number of bits for the field "HARQ process number" in DCI format 1_2 (see TS 38.212 [11], clause 7.3.1).	0...5	Extension of R16
HARQ-ProcessNumber SizeDCI-0-2-r17	Configure the number of bits for the field "HARQ process number" in DCI format 0_2 (see TS 38.212 [11], clause 7.3.1).	0...5	Extension of R16
HARQ-ProcessNumber SizeDCI-1-1-r17	Configure the number of bits for the field "HARQ process number" in DCI format 1_1.	5	New in Rel-17
HARQ-ProcessNumber SizeDCI-0-1-r17	Configure the number of bits for the field "HARQ process number" in DCI format 0_1.	5	New in Rel-17

For DCI format 1_1, HARQ process number is encoded in 5 bits if the higher layer parameter **harq-ProcessNumberSizeDCI-1-1** is configured; otherwise, 4 bits.

For DCI format 1_2, the number of bits indicating the HARQ process number is determined by the following: 0, 1, 2, 3, 4, or 5 bits determined by higher layer parameter **harq-ProcessNumberSizeDCI-1-2-v1700** if configured. Otherwise, 0, 1, 2, 3, or 4 bits are determined by higher-layer parameter harq-ProcessNumberSizeDCI-1-2.

For DCI format 0_1, HARQ process number is encoded in 5 bits if higher-layer parameter **harq-ProcessNumberSizeDCI-0-1** is configured; otherwise, 4 bits.

For DCI format 0_2, the number of bits indicating the HARQ process number is determined by the following: 5 bits determined by higher layer parameter **harq-ProcessNumberSizeDCI-0-2-v1700** if configured. Otherwise, 0, 1, 2, 3, or 4 bits are determined by higher-layer parameter harq-ProcessNumberSizeDCI-0-2.

For DCI 0-0/1-0 (also known as fallback DCI formats), it is not necessary to schedule 32 HARQ processes, therefore, no enhancement to support indication of more than 16 HARQ processes is considered in Rel-17.

The new and updated higher-layer parameters related to the enhancement of the HARQ process indication in 3GPP Release-17 are listed in Table 5.15.

Increasing the number of HARQ processes in NTN may have an impact on the HARQ soft buffer at the UE. However, this is not likely to be increased significantly considering that typically NTN has lower system bandwidth compared to TNs, especially in FR1, and MIMO with rank-1 is typically assumed in NTN systems.

5.6.3 Disabling HARQ Feedback in NTN

Solutions to avoid reduction in peak data rates in NTN were specified in Release-17. One solution is to increase the number of HARQ processes to match the longer satellite RTD to avoid HARQ stalling as described in Section 5.6.2. Another solution is to disable Hybrid-ARQ feedback to avoid stop-and-wait in HARQ procedure and rely on faster ARQ re-transmissions over the RLC layer for reliability.

5.6.3.1 Disabling HARQ Feedback Activation

In NTN, enabling and disabling HARQ feedback for downlink transmission can be semi-statically configurable per HARQ process via UE-specific RRC signaling. The parameter **downlinkHARQ-FeedbackDisabled-r17**, which is a bit string of 32 bits is introduced in Release-17 to enable or disable the HARQ-feedback per HARQ process.

In NR, Semi-Persistent Scheduling (SPS) in DL can be used, e.g., to support non-full buffer traffic such as VoNR. With SPS, a user can be allocated time-frequency resources in a semi-persistent manner, i.e., fixed resources are allocated at certain intervals without L1/L2 control signaling each time. This is especially useful to reduce the L1/L2 control signaling overhead and to increase VoIP capacity. For SPS, the Base Station can allocate downlink resources for the initial Hybrid ARQ transmissions to UE: RRC defines the periodicity of the configured downlink assignments while PDCCH scrambled by CS-RNTI can either signal and activate the configured downlink assignment, or deactivate it, i.e., a PDCCH addressed to CS-RNTI indicates that the downlink assignment can be implicitly reused according to the periodicity defined by RRC, until deactivated.

In NR NTN, for HARQ feedback of each SPS PDSCH, UE follows the per-process configuration of HARQ feedback enabled/disabled for the associated HARQ process, except for the first SPS PDSCH after activation if enabled by the network via RRC configuration using the parameter **HARQ-feedbackEnablingforSPSactive-r17**: If enabled, UE reports ACK/NACK for the first SPS PDSCH after activation, regardless of whether HARQ feedback is enabled or disabled corresponding to the first SPS PDSCH after activation. Otherwise, UE follows configuration of HARQ feedback enabled/disabled corresponding to the first SPS PDSCH after activation. Further, for DCI indicating SPS PDSCH release, Hybrid-ARQ-ACK report is as in Release-16.

Disabling HARQ feedback in NTN requires enhancements to HARQ ACK codebooks, and PDSCH/PUSCH scheduling restrictions. It may also require enhancements to improve reliability of transmission when HARQ is disabled with multiple transmissions of the same TB and soft combining. Such performance enhancements were discussed, but not agreed to be part of 3GPP Release-17.

5.6.3.2 HARQ ACK Codebook Enhancements

In 5G NR, the HARQ ACK codebook allows the UE to multiplex the HARQ ACK/NACK from multiple slots, TBs, multiple Code Block Groups (CBG), and multiple carriers, into one multi-bit acknowledgment message. The HARQ ACK codebook defines the format used to signal a set of HARQ acknowledgments. The following categories of codebook have been specified:

- Type 1 HARQ ACK codebook: semi-static size which is fixed by information provided by RRC signaling.
- Type 2 HARQ ACK codebook: dynamic size that changes according to the number of recourse allocations.
- Type 3 HARQ ACK codebook: also known as one-shot HARQ-ACK codebook feedback; the UE sends Hybrid-ARQ report for all HARQ processes, and all CCs configured in one PUCCH group.

In 3GPP Release-17, enhancements for Type-1, Type-2, and Type-3 HARQ codebooks have been introduced in order to provide feedback to the gNB for DL data transmission with feedback-disabled HARQ processes.

The principle of Type 1 HARQ ACK codebook is recalled as follows:

- The size of Type 1 HARQ ACK codebook is semi-static and equals the sum of all possible PDSCH transmission occasions within a specific time window. This sum accounts for the possibility of multiple PDSCH transmitted within a slot, PDSCH transmitted across slots, and PDSCH transmitted across carriers. It accounts also for multiple TBs belonging to a specific PDSCH and multiple CBG belonging to a specific TB.
- The time span of the codebook or the window across which the transmission occasions are summed is given by the list of timing for given PDSCH to the DL ACK, which depends on the SCS configuration of PUCCH transmission and DCI format used for PDSCH allocation, e.g., in case of DCI format 1_0, the list of slot timing is {1, 2, 3, 4, 5, 6, 7, 8} for

SCS configuration of PUCCH transmission $\mu \leq 3$. In case of DCI format 1_1, the list of slot timing is configured by **DL-DataToUL-ACK-r16** or **DL-DataToUL-ACK-r17**, or **DL-DataToUL-ACK-v1700**.

To support HARQ feedback disabling in NTN, for Type-1 HARQ codebook, the UE will consistently report NACK-only for the feedback-disabled HARQ process regardless of decoding results of corresponding PDSCH. Example of semi-static/Type-1 HARQ codebook is illustrated in Figure 5.30. In this example, the UE is configured to receive DL data in a carrier. Each transmission includes a single TB. The list of timing for given PDSCH to the DL ACK is given by **DL-DataToUL-ACK-v1700**, which equals {31, 30, 29, 22, 21, 19, 18, 16} meaning that the transmission of a HARQ ACK codebook at slot n can include acknowledgment from PDSCH transmissions, which were received during slots n-**31**, n-**30**, n-**29**, n-**22**, n-**21**, n-**19**, n-**18**, and n-**16**. There are entries within the Type 1 HARQ ACK codebook for each of these slots irrespective of whether or not any downlink data was actually transmitted. The downlink data is not transmitted during the slots indexed as −30 and −21, the UE populates these entries within the codebook using negative acknowledgments. Further, the UE is provided **downlinkHARQ-FeedbackDisabled-r17** indicating disabled HARQ-ACK information in PDSCH reception occasion indexed as −**31**, −**29**, −**22**, −**18**, and −**16**, thereby the UE populates also these entries within the codebook using negative acknowledgments. The gNB knows that these slots are associated with disabled HARQ-ACK information and so expects to receive negative acknowledgments. In this example, it is assumed that HARQ-ACK information is enabled only on slot indexed as −19. HARQ acknowledgment is generated for this slot based on the success or failure of the decoding process. If a Type-1 HARQ-ACK codebook would not include any HARQ-ACK information for TBs with enabled HARQ-ACK information, the UE does not provide the Type-1 HARQ-ACK codebook and does not transmit a corresponding PUCCH.

DL-DataToUL-ACK-v1700 = {31,30,29,22,21,19,18,16}

Slot	−31	−30	−29	...	−22	−21	−20	−19	−18	−17	−16
DL HARQ Feedback	disabled	enabled	disabled		disabled	disabled		enabled	disabled		disabled
	S	N	S		S	N		S	S		S

S= Scheduled transmission I N= Non-scheduled transmission

Slot	−31	−30	−29	...	−22	−21	−20	−19	−18	−17	−16
UE generates HARQ-ACK infromation	NACK	NACK	NACK		NACK	NACK		ACK/ NACK	NACK		NACK

Figure 5.30 Example of enhanced semi-static (Type-1) HARQ-ACK codebook.

DL-DataToUL-ACK-v1700 = {31,30,29,26,25,23,21,20}

Slot	-31	-30	-29	...	-22	-21	-24	-23	-22	-21	-20
DL HARQ Feedback	disabled	enabled	disabled		disabled	disabled		disabled		enabled	disabled
A single serving cell. A maximum of 4 CBG configured	S	S	S		S	S		N		S	S
	S	N	S		S	S		N		N	S
	S	N	S		S	S		N		N	S
	S	N	S		S	N		N		N	S

S= Scheduled transmission | N= Non-scheduled transmission

Slot	-31	-30	-29	...	-26	-25	-24	-23	-22	-21	-20
	NACK	ACK/NACK	NACK		NACK	NACK		NACK		ACK/NACK	NACK
UE generates HARQ-ACK infromation	NACK	NACK	NACK		NACK	NACK		NACK		NACK	NACK
	NACK	NACK	NACK		NACK	NACK		NACK		NACK	NACK
	NACK	NACK	NACK		NACK	NACK		NACK		NACK	NACK

Figure 5.31 Example of enhanced semi-static (Type-1) HARQ-ACK codebook and CBG configured.

In the example in Figure 5.31, the UE is also provided **PDSCH-CodeBlockGroupTransmission**, with maximum number of CBGs per TB equals 4. As illustrated in this figure, the Type 1 HARQ ACK codebook contains entries for all 4 CBGs irrespective of the actual number of CBGs, e.g., only one CBG is transmitted in slots indexed as −30 and −21 but the UE populates four entries within the cookbook for each of the two slots. The UE is provided downlinkHARQ-FeedbackDisabled-r17 indicating disabled HARQ-ACK information in PDSCH reception occasion indexed as −31, −29, −26, −25, and −20, thereby the UE populates negative acknowledgments within the cookbook entries corresponding to CBGs of the TB.

The main drawback of the semi-static codebook is the potentially large size of a hybrid-ARQ report. Type 2 HARQ ACK codebook is used to address this drawback by excluding codebook entries corresponding to unused transmission occasions and as per Release-17 enhancements, excluding entries corresponding to PDSCH with feedback-disabled HARQ processes.

The principle of Type 2 HARQ ACK codebook is recalled as follows:

- As mentioned above, with a dynamic codebook, only the acknowl-edgment information for the scheduled transmission occasions with feedback-enabled HARQ processes is included in the hybrid-ARQ report, instead of all transmission occasions, scheduled or not, with feedback-disabled HARQ processes or not, as is the case with a semi-static codebook.

- The main challenge with Type 2 HARQ ACK codebook is to maintain the correct relationship between acknowledgments and transmissions [18]. In fact, in the presence of an error in the downlink control signaling, the UE and gNB may have different understandings of the number of scheduled transmission occasions, which would lead to an incorrect codebook size and possibly corrupt the entire feedback report. In the example shown in Figure 5.23, if the UE, for example, missed the PDCCH transmitted at occasion −19, then the UE would generate a Hybrid-ARQ codebook with smaller size and the gNB would map acknowledgments onto incorrect transmissions.
- The counter DAI (cDAI) and total DAI (tDAI) derived from the DAI included in the DCI containing the downlink assignment (refer to Table 5.12) are used to avoid error cases created by missed DL control signaling. tDAI is needed in the case of carrier aggregation. If a single serving is used, then only cDAI is required.
- For the DCI of PDSCH with feedback-enabled HARQ processes, the C-DAI and T-DAI are the counts of only feedback-enabled processes.

An example of enhanced dynamic (Type-2) HARQ-ACK codebook is shown in Figure 5.32: only the entries with feedback-enabled HARQ processes are included in the hybrid-ARQ report and the other entries (which correspond to PDSCH with feedback-disabled HARQ processes or non-scheduled remissions) are omitted. This reduces the size of the acknowledgment message. This example illustrates also the principle of using the counter DAI. If the UE receives non-consecutive counter value such as on occasion indexed −18, the UE detects a missed transmission. Thereby, the UE can insert a negative acknowledgment within the HARQ

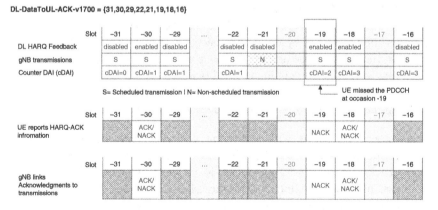

DL-DataToUL-ACK-v1700 = {31,30,29,22,21,19,18,16}

Slot	−31	−30	−29	...	−22	−21	−20	−19	−18	−17	−16
DL HARQ Feedback	disabled	enabled	disabled		disabled	disabled		enabled	enabled		disabled
gNB transmissions	S	S	S		S	N		S	S		S
Counter DAI (cDAI)	cDAI=0	cDAI=1	cDAI=1		cDAI=1			cDAI=2	cDAI=3		cDAI=3

S= Scheduled transmission | N= Non-scheduled transmission — UE missed the PDCCH at occasion -19

Slot	−31	−30	−29	...	−22	−21	−20	−19	−18	−17	−16
UE reports HARQ-ACK infromation		ACK/NACK						NACK	ACK/NACK		

Slot	−31	−30	−29	...	−22	−21	−20	−19	−18	−17	−16
gNB links Acknowledgments to transmissions		ACK/NACK						NACK	ACK/NACK		

Figure 5.32 Example of enhanced dynamic (Type-2) HARQ-ACK codebook.

report (bit corresponding to occasion −19 in the report) which allows to maintain the relationship between acknowledgments and transmissions.

Type 3 HARQ ACK codebook or one-shot HARQ-ACK codebook feedback was introduced in Release-16 to the NR-based access to NR-U system. includes feedback information corresponding to all HARQ processes on carriers configured in one PUCCH group. The Base Station can configure one-shot HARQ-ACK codebook feedback for the UE **pdsch-HARQ-ACK-OneShotFeedback-r16** or **pdsch-HARQ-ACK-OneShotFeedbackDCI-1-2-r17** and trigger the UE to operate accordingly via the DCI format 1_1 or DCI format 1_2.

For Type-3 HARQ codebook in NTN, the UE should skip the codebook feedback for feedback-disabled HARQ processes. The Type-3 codebook size is reduced by excluding the bit positions of disabled HARQ processes. Type-3 HARQ-ACK codebook determination is specified in clause 9.1.4 of TS 38.213 [6].

5.6.4 Transmission Timing for HARQ-ACK on PUCCH

5G NR supports asynchronous and adaptive Hybrid ARQ is in the downlink. Asynchronous HARQ means that there is no fixed timing pattern for each HARQ process. The Base Station provides the UE with the HARQ-ACK feedback timing either dynamically in the DCI or semi-statically in an RRC configuration. Timing between DL data reception and corresponding acknowledgment is indicated by PDSCH-to-HARQ feedback timing indicator field in the DCI. Specifically, this field determines the delay between the end of the slot carrying the PDSCH and the start of the slot used to return the HARQ ACK/NACK.

5.6.4.1 Transmission Timing for HARQ-ACK on PUCCH in Terrestrial Network

In NR TN, the transmission timing for HARQ-ACK on PUCCH relative to the reception of the PDSCH is defined as follows: For a PDSCH reception ending in slot n, the UE reports corresponding HARQ-ACK information in a PUCCH transmission within slot $n + K_1$, where K_1 is a number of slots and is indicated by the PDSCH-to-HARQ-timing-indicator field (PHFTI) in the DCI. Depending on the DCI format used for PDSCH resource allocation, PHFTI maps into an RRC-configured look-up table or directly maps into a predefined set of values:

- In case of DCI format 1_0, PHFTI maps into {1, 2, 3, 4, 5, 6, 7, 8} for SCS configuration of PUCCH transmission $\mu \leq 3$. Other predefined set of values are specified for higher numerologies (refer to clause 9.2.3 in TS 38.213 [6]).

- In case of DCI format 1_1 and DCI format 1_2, PHFTI is used as an index into RRC-configured look-up table provided by **dl-DataToUL-ACK** which applies to DCI format 1_1, **dl-DataToUL-ACK-DCI-1-2** which applies to DCI format 1_2, the **dl-DataToUL-ACK-v1700** which is applicable for NTN and **dl-DataToUL-ACK-r17** which is applicable for up to 71 GHz. If **dl-DataToUL-ACK-r16** or **dl-DataToUL-ACK-r17** or **dl-DataToUL-ACK-v1700** is signaled, UE shall ignore the **dl-DataToUL-ACK** (without suffix).

The mapping of PHFTI field values to numbers of slots is defined by table 9.2.3-1 of TS 38.213 [6].

In particular, K_1 can be set equal to 0 which corresponds to a self-contained slot, where the delay from the end of the data transmission to the transmission of the acknowledgment from the UE is within a slot.

Figure 5.33 illustrates an example of PDSCH-to-HARQ feedback timing indicated in DCI 1_1: in this example, in DL slot n, PHFTI = 0, which is pointing into $k1$ value = 2, in related RRC Table provided by **dl-DataToUL-ACK**. Thereby, slot $n + 2$ is the slot used to return the HARQ ACK/NACK corresponding to PDSCH transmitted in slot n.

5.6.4.2 Transmission Timing for HARQ-ACK on PUCCH in NTN

In NR NTN, the timing of the PUCCH carrying the HARQ-ACK information, is defined by the assigned HARQ-ACK timing K1 (as described in Section 5.6.4.1) and Koffset, if configured. Specifically, for the transmission

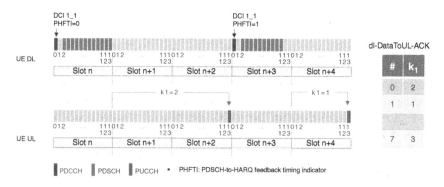

Figure 5.33 Transmission timing for HARQ-ACK on PUCCH in terrestrial network.

timing of HARQ-ACK on PUCCH, the UE provides corresponding HARQ-ACK information in a PUCCH transmission within slot:

$$n + K_1 + K_{offset} \cdot \frac{2^\mu}{2^{\mu_{K_{offset}}}}$$

where

- $K_{offset} = K_{cell,offset} - K_{UE,offset}$, where $K_{cell,offset}$ is provided by **CellSpecificKoffset** in **ServingCellConfigCommon** and $K_{UE,offset}$ is provided by a MAC CE command; otherwise, if not respectively provided, $K_{cell,offset} = 0$ or $K_{UE,offset} = 0$.
- μ is the SCS configuration for the PUCCH transmission.
- $\mu_{K_{offset}}$ is the SCS configuration for K_{offset} with a value of 0 for frequency range, FR1.
- $K1$ is provided by the PHFTI field in the DCI as described in Section 5.6.4.1.

Figure 5.34 illustrates an example of the transmission timing for HARQ-ACK on PUCCH in NTN: in this example, in DL slot n, PHFTI = 0 which points into $k1$ value = 2, in related RRC Table provided by dl-DataToUL-ACK-v1700. Thereby, slot $n + 2 + K\_offset$ is the slot used to return the HARQ ACK/NACK corresponding to PDSCH transmitted in slot n.

5.6.4.3 PDSCH Scheduling Restriction

UE PDSCH processing time is specified in TS 38.214 [20]. The PDSCH processing capability T_proc,1 defines the minimum time that an UE requires between the end of PDSCH reception and the first uplink symbol of the PUCCH which carries the HARQ-ACK information. Two categories of PDSCH processing capability have been defined in clause 5.3 of TS 38.214 [20].

In essence, the UE is not expected to receive a PDSCH that overlaps in time with another PDSCH for any HARQ process: When HARQ feedback for the HARQ process ID is enabled, the UE is not expected to receive another PDSCH for a given HARQ process until after the end of the expected transmission of HARQ-ACK for that HARQ process. The transmission timing for HARQ-ACK is determined as described in Section 5.6.4.2 and includes the UE PDSCH processing time.

For a DL HARQ process with disabled HARQ feedback, the UE is not expected to receive another PDSCH or set of slot-aggregated PDSCH scheduled for the given HARQ process or to receive another PDSCH without

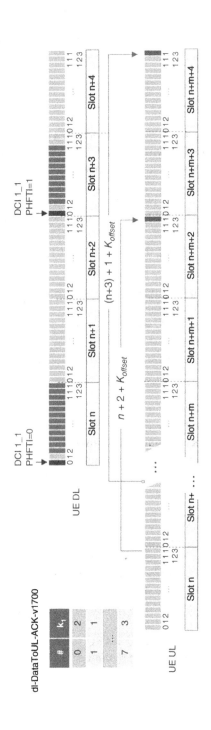

Figure 5.34 Transmission timing for HARQ-ACK on PUCCH in NTN.

corresponding PDCCH for the given HARQ process that starts until Tproc,1 after the end of the reception of the last PDSCH or slot-aggregated PDSCH for that HARQ process.

References

1 3GPP TR 38.811: "Study on New Radio (NR) to support non-terrestrial networks," v15.4.0, October 2020.

2 3GPP TR 38.821: "Solutions for NR to support non-terrestrial networks (NTN)," v16.0.0, January 2020.

3 Introduction to the 3GPP defined NTN standard – A comprehensive view on the 3GPP work on NTN, M. El Jaafari et al, IJSCN, February 2023.

4 3GPP TS 38.300: "NR and NG-RAN Overall Description; Stage 2."

5 3GPP TS 38.211: "Physical channels and modulation (Release 17)," March 2022.

6 3GPP TS 38.213: "Physical layer procedures for control (Release 17)," December 2022.

7 3GPP TS 38.104: "Base Station (BS) radio transmission and reception (Release 18)," December 2022.

8 3GPP TS 38.108: "Satellite Access Node radio transmission and reception."

9 RP-230809, Revised WID: NR NTN (Non-Terrestrial Networks) enhancements, Thales, 3GPP TSG RAN Meeting #99, March 2023.

10 R1-2005311, THALES, Considerations on PAPR requirements for NR NTN downlink transmission, RAN1#102e, August 2020.

11 3GPP TS 38.212: "Multiplexing and channel coding (Release 17)," December 2022.

12 3GPP TS 38.306: "User Equipment (UE) radio access capabilities (Release 17)," September 2022.

13 Physical layer enhancements in 5G-NR for direct access via satellite systems, S. Cioni et al., IJSCN, July 2022.

14 3GPP TS 38.133: "Requirements for support of radio resource management."

15 3GPP TS 38.331: "Radio Resource Control (RRC) protocol specification (Release 17)," December 2022.

16 R1-2201011, Maintenance on UL timing and frequency synchronization in NTN Thales, March 2022.

17 R1-2111122, Considerations on UL timing and frequency synchronization in NTN Thales, November 2021.

18 *5G NR the Next Generation Wireless Access Technology*, 2nd ed.
E. Dahlman, S. Parkvall, J.Sköld, Elsevier Science, 2020

19 3GPP TS 38.101-5: "User Equipment (UE) radio transmission and reception, Part 5: Satellite access Radio Frequency (RF) and performance requirements."

20 3GPP TS 38.214: "Physical layer procedures for data (Release 17)," December 2022.

21 3GPP TS 38.321: "Medium Access Control (MAC) protocol specification."

6

Impacts on the System Architecture and Network Protocol Aspects

6.1 Introduction

In Chapters 2, 3, 4, and 5, we have introduced the basic Non-Terrestrial Networks (NTN) concepts and the standardized architecture building blocks. In this chapter, we will describe the standardized NTN functionalities, related to the impact on system architecture and network protocols. In more detail, Quality of Service (QoS) handling, UE attach, mobility, feeder link switch, network, and radio resource management, and the procedures up to the Radio Link Control (RLC) and Packet Data Convergence Protocol (PDCP) layers are described. We will focus on 3rd Generation Partnership Program (3GPP) Rel-17, but we will also mention some ongoing enhancements currently discussed in Rel-18.

6.2 5G QoS and NTN

In general, provisioning of services is based on their assigned QoS. This reflects the fact that different services have different requirements on the bit rate, delay, jitter, and so on; furthermore, since radio and transport network resources are limited and many users may share the same available resources, efficient mechanisms are available to partition such resources among the applications and the users [1]. The 5th Generation Core Network (5GC) and the New (G) RAN (NG-RAN) need to ensure that the different service requirements are supported and that the services receive the appropriate QoS treatment to enable the expected user experience.

A detailed description of QoS treatment in 5G is out of the scope of this book. We will present some definitions and concepts, which are important to understand the impact of NTN on NG-RAN and 5GC.

5G Non-Terrestrial Networks: Technologies, Standards, and System Design, First Edition.
Alessandro Vanelli-Coralli, Nicolas Chuberre, Gino Masini, Alessandro Guidotti, and Mohamed El Jaafari.

The 5G QoS model is based on the concepts of Packet Data Unit *(PDU) Sessions* and *QoS Flows*. A PDU Session is the association between the User Equipment (UE) and a data network providing a PDU connectivity services, and the QoS flow is the finest granularity of QoS differentiation within a PDU Session. Each QoS Flow is identified by a QoS Flow ID (QFI) and is assigned a QoS profile, one or more QoS rules, and one or more Packet Detection Rules (PDRs) [2].

The QoS profile includes the QoS parameters: 5G QoS Identifier (5QI) and Allocation and Retention Priority (ARP). The 5QI identifies a specific forwarding behavior (packet loss rate, packet delay budget); this may be implemented in the access network via, e.g., different scheduling weights, admission thresholds, queue management thresholds, link layer protocol configuration, and so on. The ARP indicates a priority level for the QoS flow allocation and retention: it is used by the network to decide whether to accept or reject a QoS flow when resources are limited, and to prioritize such requests [1, 2]. QoS rules are used by the UE to associate UpLink (UL) traffic with QoS flows. PDRs are used in the core network to classify a packet arriving at a User Plane Function (UPF).

The principle for User Plane (UP) traffic classification and marking, and mapping of QoS Flows to access network resources, is shown in Figure 6.1 [2], for convenience. This is the conceptual model behind the handling of data packets from applications by the 3GPP system.

A Policy and Charging Control (PCC) rule contains a set of information (in a Service Data Flow [SDF] template) used to identify the IP packets that

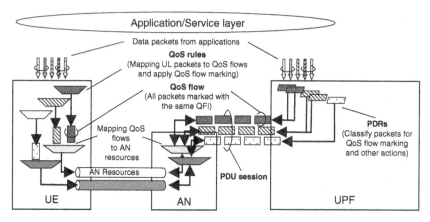

Figure 6.1 Classification and UP marking for QoS flows, and mapping to AN resources. Source: 3GPP TS 23.501 [2]/ETSI.

belong to a service session. An SDF template contains one or more packet filters; all IP packets matching these packet filters are designated as an SDF. Each filter contains a description of the IP flow, and typically contains the source and destination addresses and ports, the protocol type (often referred to as the IP 5-tuple). The PCC rule also contains the gating status (open/closed), as well as QoS and charging-related information for the SDF [1]. The system maps the PCC rules to QoS Flows based on the QoS and service requirements; a QFI for a new QoS Flow is assigned, and its QoS profile, corresponding UPF instructions, and QoS rules are derived [2].

Standardized 5QI values are specified for services that are assumed to be frequently used, thus benefiting from optimized signaling. Dynamically assigned 5QI values (which require signaling the QoS characteristics as part of the QoS profile) can be used for services for which a standardized 5QI is not defined [2].

The complete mapping of standardized 5QI values to the various QoS characteristics as defined by 3GPP is shown in Annex (see Table 6.A.1), which also mentions some example services. Notice that at least for one 5QI (5QI 10), the standard has made provisions for propagation delays, which may be encountered with LEO and GEO satellite systems (Table 6.A.1, refer footnote *q*).

6.3 Network Attach, AMF Selection, and UE Location

6.3.1 Network Identities

An NTN is shown in Figure 4.1. To provide service to a UE within the 3GPP system, such an NTN supports the NG-RAN network identities as defined by 3GPP. These identities are specified in [4] and are listed below for convenience [4]:

- Access and Mobility Function (AMF) Name: Used to identify an AMF.
- NR Cell Global Identifier (NCGI): Used to identify NR cells globally. The NCGI is constructed from the Public Land Mobile Network (PLMN) identity the cell belongs to and the NR Cell Identity (NCI) of the cell. The PLMN ID included in the NCGI should be the first PLMN ID within the set of PLMN IDs associated with the NR Cell Identity in SIB1, following the order of broadcast.
- gNB Identifier (gNB ID): Used to identify gNBs within a PLMN. The gNB ID is contained within the NCI of its cells.

- Global gNB ID: Used to identify gNBs globally. The Global gNB ID is constructed from the PLMN identity the gNB belongs to and the gNB ID. The MCC and MNC are the same as included in the NCGI.
- Tracking area identity (TAI): Used to identify tracking areas (TAs). The TAI is constructed from the PLMN identity the TA belongs to and the Tracking area code (TAC) of the TA. For NTN, a TA corresponds to a *fixed geographical area*; any required mapping is configured in the RAN. Multiple TACs may be broadcasted in an NTN cell.
- Single network slice selection assistance information (S-NSSAI): Identifies a network slice.
- Network Slice AS Group (NSAG): Identifies a slice or a set of slices. An NSAG is defined within a TA, used for slice-specific cell reselection and/or slice-specific RACH configuration.
- Network Identifier (NID): Identifies an SNPN in combination with a PLMN ID.
- Closed Access Group (CAG) Identifier: Identifies a CAG within a PLMN.
- Local NG-RAN Node Identifier: Used as reference to the NG-RAN node in the Inactive state Radio Network Temporary Identifier (I-RNTI).
- Mapped Cell ID: The Cell ID indicated to the 5GC as part of the User Location Information (ULI), and used for paging optimization over the NG interface, for Areas of Interest (AoI), and Public Warning Service (PWS), corresponds to a mapped Cell ID. The Cell ID included in messages exchanged over network interfaces allows identifying the correct cell. The mapping between Mapped Cell IDs and geographical areas is configured in the Radio Access Network (RAN) and in the core network; this allows correct support of fixed and moving cells with LEO constellations.

In association with a TAC, at NG interface setup the NTN gNB signals to the core network whether it supports a LEO, a MEO, a GEO, or "other" type of NTN (Radio Access Technology [RAT] Type). When the UE is accessing NR using satellite access, this type of access is indicated to the core network [2, 5]. This is also used to enforce any mobility restrictions. A RAT Type is expected to be homogeneous within a TA (and is not expected to be mixed with a terrestrial access RAT in the same TA) [2].

6.3.2 Multiple TACs Support

Each TA is assumed to remain fixed with respect to the Earth even if the radio cells are moving across the Earth's surface. To achieve this, the NG-RAN may change the broadcasted TACs as the cell moves. In other words, the RAN node may broadcast in a cell a single TAC per PLMN and change that TAC as the cell moves. The NG-RAN may also broadcast more than one TAC for a PLMN and add or remove TACs as the cell moves. The NG-RAN provides either the single broadcast TAI or all broadcast TAIs corresponding to the Selected PLMN to the AMF as part of the ULI. The NG-RAN indicates, if known, also the TAI where the UE is geographically located [2, 5].

6.3.3 UE Attach and Location Verification

The UE can determine the network type (terrestrial or NTN) by looking at the information broadcasted in the SIB by each cell.[1] When a UE attaches to the network (registration procedures are the same as for 5G terrestrial networks and are specified in [6]), an appropriate AMF for that UE must be selected,[2] to enable proper routing via the NG interface. This function (Non-Access Stratum, NAS Node Selection Function, NNSF) resides in the NG-RAN node to determine the AMF association for the UE, based on, e.g., the temporary identifier for the UE, slicing information, onboarding information, serving cell, and estimated UE location [2, 7].

For regulatory/licensing purposes, the network may be configured to enforce the operation of a certain PLMN in a certain geographical area.[3] In this case, the UE location needs to be verified by the network during mobility management and session management procedures [2]. If the gNB detects that the UE is in a different country than the one served by the selected AMF, then it should perform an NG handover to change to an appropriate AMF or initiate a UE context release request procedure toward the serving AMF (which in turn may decide to de-register the UE) [2, 4, 5].

While a UE may be able to securely provide to the serving gNB its own location (e.g., obtained by its own GPS receiver), such information is generally not considered trusted.[4] For this reason, it is required that the network verifies the UE location autonomously.

---

1 This information includes the satellite ephemeris for the serving cell and for the neighbor cells.
2 The interconnection of NG-RAN nodes to multiple AMFs is supported in the 5G architecture.
3 This may be necessary to route traffic, support emergency calls, and provide appropriate services according to the applicable national or regional regulations. Even when providing services over entire continents with NTN, there is currently no globally harmonized set of requirements that overrules local ones (similarly to terrestrial networks).
4 GPS spoofing is a common attack scenario where an external radio transmitter could be used to send a counterfeit GPS signal to the UE to counter the legitimate GPS signal, causing an incorrect location to be reported.

If the deployment of the NTN cells is granular enough, the serving cell ID known by the serving gNB is sufficient to determine the UE location accurately enough for this purpose (5–10 km accuracy is considered sufficient to determine the country where the UE is currently located) [8]. However, for deployments of large NTN cells, possibly covering more than one country (in which case different core networks for different countries may be connected to the same NTN RAN), the cell ID alone may not be sufficient for this purpose. In this case, in Rel-18 it is expected that the (temporary) serving AMF initiates a location services request toward the Location Management Function (LMF) in the core network. The LMF in turn initiates location procedures with the serving, and possibly neighbor, gNBs, to obtain positioning measurements or assistance data, then providing a UE location estimate to the AMF [9].

6.4 Random-access Procedure

The random-access procedure is triggered by a number of events, such as initial access from RRC_IDLE, RRC Connection Re-establishment procedure, and Handover. Random-access procedure is also needed in the following cases: DL or UL data arrival during RRC_CONNECTED when UL synchronization status is "non-synchronized," transition from RRC_INACTIVE, request for Other System info (SI) and during different types of protocol failures such as beam failure recovery.

The random-access procedure can be contention-based, e.g., at initial connection from idle mode) or non-contention based, e.g., during Handover to a new cell. Random-access resources and parameters are configured by the network and signaled to the UE via broadcast or dedicated signaling.

As illustrated in Figure 6.2, contention-based random access consists of four steps, usually termed message 1 (Msg1) to message 4 (Msg4): It encompasses the transmission of a random-access preamble (Msg1) by the UE using a PRACH subject to possible contention with other UEs, followed by a random access response (RAR) in DL, including allocating specific radio resources for the uplink transmission. Afterward, the UE transmits the initial UL message, e.g., Radio Resource Control (RRC) connection Request using the allocated resources and waits for a contention resolution message in DL confirming access to that UE. The UE could perform multiple attempts until it is successful in accessing the channel or until a timer supervising the procedure elapses.

Non-contention–based random access procedure foresees the assignment of a dedicated random-access resource or preamble to a UE, e.g., part of

Figure 6.2 Random Access Procedure.

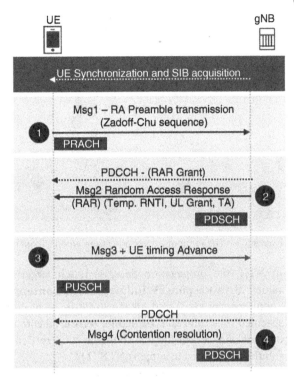

an HO command. This avoids the contention resolution phase, i.e., only the random-access preamble and RAR messages are needed to get channel access.

The random-access procedure has defined monitoring windows maintained by the UE for Msg2 and Msg4. This is to allow for power savings and reduce the risk of accidentally receiving the wrong response: After transmitting the Random-Access Preamble (Msg1), the UE monitors the PDCCH for the RAR message (Msg2). The response window (**ra-ResponseWindow**) starts at a determined time interval after the preamble transmission. In terrestrial communications, the RAR is expected to be received by the UE within a few milliseconds after the transmission of the corresponding preamble. The network configures a value lower than or equal to 10 ms when Msg2 is transmitted in licensed spectrum and a value lower than or equal to 40 ms when Msg2 is transmitted with shared spectrum channel access [10]. When the UE sends an RRC Connection Request (Msg3), it will monitor for Msg4 in order to resolve a possible random-access contention. The ra-ContentionResolutionTimer starts after Msg3 transmission. The value of the contention resolution timer could be up to 64 subframes [10].

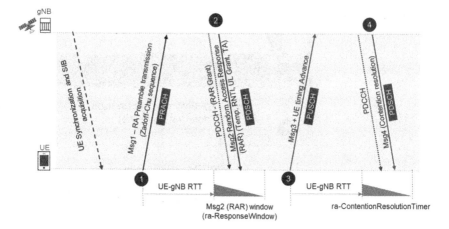

Figure 6.3 The random-access procedure when UE operates in NTN.

In NTN, the propagation delay is much larger than the configurable values of **ra-ResponseWindow** and **ra-ContentionResolutionTimer**. Therefore, the Msg2 and Msg4 cannot be received by the UE within the specified time interval specified for terrestrial communications. Therefore, the behavior of *ra-ResponseWindow* and *ra-ContentionResolutionTimer* should be modified to support NTN [11].

As specified in 3GPP Release-17 and illustrated in Figure 6.3, for the random-access procedure when UE operates in NTN, to ensure that the correct window is monitored, the UE shall support estimating UE-gNB RTT[5] [12]; corresponding to the RTT between the UE and the base station, and delaying the start of RAR window and the start of ra-ContentionResolutionTimer by UE-gNB RTT.

For the gNB to be aware of the value of estimated propagation delay between the UE and the base station, UE reporting of information related to the timing advance through a MAC Control Element was introduced for NTN as specified in [13]. It is mandatory to support TA reporting during initial access for UEs supporting **uplink-TA-Reporting-r17** as specified in [13]. Timing Advance reporting is controlled by RRC by configuring the following two parameters: **offsetThresholdTA** and **timingAdvanceSR**. It is triggered upon indication from upper layers to trigger a Timing Advance report, or upon configuration of **offsetThresholdTA** by upper layers, if the UE has not previously reported Timing Advance value to current Serving Cell, or if the variation between the current estimate of the Timing Advance

---

5 UE-gNB RTT: For non-terrestrial networks, the sum of the UE's Timing Advance value (see TS 38.211 clause 4.3.1) and kmac provided in NTN-config.

value and the last reported Timing Advance value equals or larger than **offsetThresholdTA**, if configured. The parameter **ta-Report** is used to indicate whether Timing Advance reporting during Random-Access procedure is enabled. The parameter **timingAdvanceSR** is used to configure whether a Timing Advance report may trigger a Scheduling Request (SR) as specified in [13].

6.5 Other Enhancements at MAC

3GPP Release-17 specified other MAC enhancements to adapt existing mechanisms and procedures to the NTN context. These include enhancements on Hybrid automatic repeat request (HARQ), Logical Channel Prioritization (LCP improvements, enhancements on Discontinuous Reception (DRX) functionality, and other improvements that will be discussed in the following subsections.

6.5.1 Hybrid ARQ Operation Enhancements

HARQ stalling due to large round-trip delay can severely impact the peak throughput in NTN. Therefore, two features have been introduced for NR NTN: HARQ feedback can be enabled or disabled dynamically per HARQ process in the presence of ARQ retransmissions at the RLC layer (e.g., in Geostationary Orbit (GSO) satellite systems) and maximum number of HARQ processes is extended to 32 (e.g., in Non-Geostationary Orbit (NGSO) satellite systems). Increasing the number of HARQ processes to 32 is sufficient in LEO-based NTN scenarios where propagation delay is moderate and HARQ can be maintained. Dynamically disabling the HARQ feedback is beneficial in scenarios where the propagation delay is longer, such as GEO scenarios where 32 HARQ processes are not sufficient to prevent HARQ stalling.

Figure 6.4 illustrates HARQ procedure where the feedback is disabled per HARQ process. HARQ feedback can be enabled or disabled in Rel-17 NTN, but HARQ processes remain configured. The criteria and decision to enable or disable HARQ feedback is under network control and is signaled to the UE via RRC in a semi-static manner: In the downlink, HARQ is disabled by an indication to not send HARQ feedback (e.g., HARQ process 3–16 in Figure 6.4). While in the uplink, there is no HARQ feedback and the gNB can schedule a new transmission or a re-transmission for a specific HARQ process.

More details on HARQ operation in NTN could be found in Chapter 5.

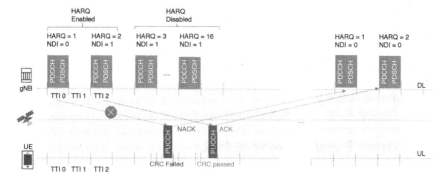

Figure 6.4 HARQ procedure in NTN.

6.5.2 Logical Channel Prioritization Enhancements

In 5G NR, MAC supports multiple numerologies and multiple MAC Service Data Units (SDUs) from the same logical channel without concatenation in RLC as shown in Figure 6.5.

LCP is a procedure in UE MAC.[6] It is used to control the multiplexing of different radio bearers associated with a MAC entity, such as signaling

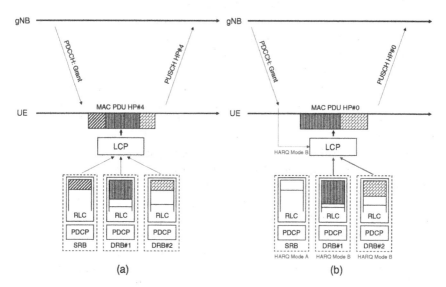

Figure 6.5 Example of logical channel prioritization (LCP) operation. Source: Määttänen et al. [14]/John Wiley & Sons.

---

6 LCP is specified for the UE in the uplink while the gNB logical channel prioritization is up to gNB implementation.

radio bearers (SRBs) containing control plane data and data radio bearer (DRBs) containing user plane data as illustrated in Figure 6.5. Further, mapping restrictions in LCP control which numerology(ies), cell(s), and transmission timing(s) a logical channel can use. The LCP should ensure that high-priority data such as control plane data is prioritized over lower-priority data such as the user plane data but also ensure that no radio bearer is starved.

One issue that was identified in NTN with the LCP procedure is that while the base station controls the prioritization through the configured LCP parameters, the base station cannot control which bearer is sent in an individual MAC PDU. With HARQ potentially providing different reliability through dynamically enabling and disabling HARQ feedback, it was considered important to make the LCP aware of the reliability provided by a specific HARQ process. Restrictions on the LCP were introduced by configuring each HARQ process to either be HARQ mode A or B and each logical channel with an allowed HARQ mode [14]. If a grant with a HARQ process is configured to be HARQ mode A or B, then only a logical channel allowed for this HARQ mode shall be included in the uplink MAC PDU. This can, for instance, ensure that control plane data sent on SRBs are not multiplexed with HARQ processes for which retransmissions are not used and thereby, control plane data is transmitted with high reliability.

The parameter **uplinkHARQ-mode** is used to set the HARQ mode (*HARQmodeA* or *HARQmodeB)* per uplink HARQ process ID [12, TS 38.321]. The parameter **allowedHARQ-mode**, if configured, includes the HARQ mode for the HARQ process associated with the UL grant. If a logical channel is configured with **allowedHARQ-mode** (*HARQmodeA* or *HARQmodeB*), it can only be mapped to a HARQ process with the same HARQ mode.

The LCP restrictions are illustrated in Figure 6.5: (a) Normal LCP operation with three bearers being multiplexed in a single MAC PDU. (b) LCP operation with restrictions based on the configured HARQ mode for the HARQ process ID.

6.5.3   Enhancements on DRX Functionality

The MAC entity may be configured by RRC with a Discontinuous Reception (DRX) functionality that controls the UE's PDCCH monitoring activity. For UE power-saving purposes, the UE may not monitor the PDCCH when no transmission is expected. Several RRC configurable parameters are used to configure DRX [10].

drx-HARQ-RTT-TimerDL[7] is the minimum duration before a downlink assignment for HARQ retransmission is expected by the MAC entity. In terrestrial communications, this is configurable in the range of a few ms, which is too small for a Satellite communication link. **drx-HARQ-RTT-TimerUL** is a similar parameter used for the uplink.

For NTN support, the DRX timers **drx-HARQ-RTT-TimerDL** and **drx-HARQ-RTT-TimerUL** have been extended by the UE-gNB RTT. Thereby, the following MAC timers are used for DRX operation in an NTN [12, TS 38.321]:

- **HARQ-RTT-TimerDL-NTN** (per DL HARQ process configured with HARQ feedback enabled): The minimum duration before a DL assignment for HARQ retransmission is expected by the MAC entity.
- **HARQ-RTT-TimerUL-NTN** (per UL HARQ process configured with HARQModeA): The minimum duration before a UL HARQ retransmission grant is expected by the MAC entity.

When DRX is configured, and a MAC PDU is received in a configured downlink assignment, if the corresponding HARQ process is configured with HARQ feedback enabled, the MAC entity shall set *HARQ-RTT-TimerDL-NTN* for the corresponding HARQ process equals *drx-HARQ-RTT-TimerDL* plus the latest available UE-gNB RTT value.

Similarly for the uplink, when DRX is configured, and a MAC PDU is transmitted in a configured uplink grant, if this Serving Cell is configured with uplinkHARQ-Mode and the corresponding HARQ process is configured as HARQModeA, the MAC entity shall set **HARQ-RTT-TimerUL-NTN** for the corresponding HARQ process equals **drx-HARQ-RTT-TimerUL** plus the latest available UE-gNB RTT value.

6.5.4 Extension of Other MAC Timers

A UE can use a SR to request UL-SCH resources from the base station for a new transmission or a transmission with a higher priority. When an SR is sent, the timer **sr-ProhibitTimer** is started. During the time the prohibit timer is active, no further SR is initiated. Prohibit time should cover round trip delay (RTD). The expiration time of **sr-ProhibitTimer** can be configured for terrestrial networks in the range between 1 and 128 ms. For GEO systems, this value range is not sufficient because the RTD is larger. Thereby, to accommodate the long propagation delay, a new timer

---

7 Value in number of symbols of the BWP where the transport block was received: INTEGER (0.56).

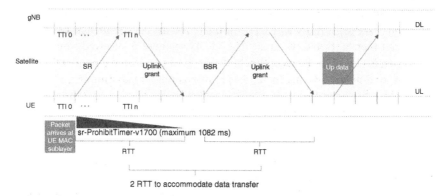

Figure 6.6 Timer for scheduling request transmission.

for SR transmission on Physical Uplink Control Channel (PUCCH) with an extended maximum value of up to 1082 ms was specified in Release-17. It is referred to as **sr-ProhibitTimer-v1700** in TS 38.331. UE shall ignore legacy timer for SR transmission on PUCCH when **sr-ProhibitTimer-v1700** is configured. The SR transmission controlled by **sr-ProhibitTimer-v1700** timer is shown in Figure 6.6.

6.6 RLC, PDCP Enhancements

Besides the extension of MAC timers discussed in Sections 6.5.3 and 6.5.4, the following RLC and PDCP sublayer timers are extended to accommodate the long propagation delay: t-Reassembly, discardTimer, and t-reordering. These timers are further discussed in the following subsections:

6.6.1 RLC Sublayer

Similar to Physical and MAC sublayers, the RLC has also to cope with the major challenge of larger propagation delay in NTN.

An AM RLC entity sends STATUS PDUs to its peer AM RLC entity in order to provide positive and/or negative acknowledgments of RLC SDUs (or portions of them). The Automatic Repeat reQuest (ARQ) within the RLC sublayer has the following characteristics:

- ARQ retransmits RLC SDUs or RLC SDU segments based on RLC status reports;
- Polling for RLC status report is used when needed by RLC;
- RLC receiver can also trigger RLC status report after detecting a missing RLC SDU or RLC SDU segment.

Detection of reception failure of an AMD PDU is indicated by the expiration of **t-Reassembly** with a maximum of 200 ms. As specified in TS 38.322, this timer is started when an AMD PDU segment is received from lower layer, is placed in the reception buffer, at least one-byte segment of the corresponding SDU is missing and the timer is not already running. The procedure to detect loss of RLC Pdus at lower layers by expiration of timer **t-Reassembly** is used in RLC AM as well as in RLC UM.

In NTN, with enabled HARQ feedback on the MAC sublayer, an extension of **t-Reassembly** is beneficial because a running **t-Reassembly** timer also includes HARQ retransmissions. Thereby, 3GPP Release-17 introduced a new **t-ReassemblyExt** with a maximum of 2200 ms [10]. If **t-ReassemblyExt-r17** is configured, the UE shall ignore **t-Reassembly** (without suffix).

6.6.2 PDCP Sublayer

The PDCP (specified in TS 38.323) performs header compression, security operations (Ciphering and Integrity protection), and guarantees in-order delivery without duplicates. The transmitting PDCP entity shall discard the PDCP SDU when the **discardTimer** expires for a PDCP SDU or when a status report confirms the successful delivery. As defined in TS 38.331, the **discardTimer** can be configured with a maximum value of 1500 ms or can be switched off by choosing infinity. To accommodate the long propagation delay in NTN, Release-17 specified the new parameter **DiscardTimerExt2-r17** with a maximum value of 2000 ms. If the UE is configured with this new parameter, the **discardTimer** and **discardTimerExt** are ignored and **discardTimerExt2** is used instead.

6.7 NTN-specific System Information

A new System Information Block (i.e., SIB19) is introduced for NTN. It contains NTN-specific parameters for serving cells and/or neighbor cells as defined in TS 38.331. The NTN UE in RRC_IDLE and RRC_INACTIVE shall ensure having a valid version of (at least) the MIB, SIB1 through SIB4, and SIB19 if UE is accessing NR via NTN access. The SIB19 is essential for NTN access. If UE is unable to acquire the SIB19 for NTN access, the action is up to UE implementation (e.g., cell re-selection to other cells). The UE in RRC_CONNECTED should attempt to re-acquire SIB19 before the end of the duration indicated by **ntn-UlSyncValidityDuration** and **epochTime** by UE implementation.

The content of SIB19 is displayed in Table 6.1:

Table 6.1 SIB19 content.

ntn-Config	This field is mandatory present for the serving cell in SIB19. It provides parameters needed for the UE to access NR via NTN access such as Ephemeris data, common TA parameters, k_offset, validity duration for UL sync information, and epoch time.
ntn-NeighCellConfigList, ntn-NeighCellConfigListExt	Provides a list of NTN neighbor cells including, their ntn-Config, carrier frequency, and PhysCellId. This set includes all elements of ntn-NeighCellConfigList and all elements of ntn-NeighCellConfigListExt. If ntn-Config is absent for an entry in ntn-NeighCellConfigListExt, the ntn-Config provided in the entry at the same position in ntn-NeighCellConfigList applies. Network provides ntn-Config for the first entry of ntn-NeighCellConfigList. If the ntn-Config is absent for any other entry in ntn-NeighCellConfigList, the ntn-Config provided in the previous entry in ntn-NeighCellConfigList applies.
distanceThresh	Distance from the serving cell reference location is used in location-based measurement initiation in RRC_IDLE and RRC_INACTIVE, as defined in TS 38.304 [15]. Each step represents 50 m.
referenceLocation	Reference location of the serving cell provided via NTN quasi-Earth fixed system and is used in location-based measurement initiation in RRC_IDLE and RRC_INACTIVE, as defined in TS 38.304.
t-Service	Indicates the time information on when a cell provided via NTN quasi-Earth fixed system is going to stop serving the area it is currently covering. The field indicates a time in multiples of 10 ms after 00:00:00 on Gregorian calendar date 1 January 1900 (midnight between Sunday, 31 December 1899 and Monday, 1 January 1900). The exact stop time is between the time indicated by the value of this field minus 1 and the time indicated by the value of this field.

6.8 Mobility Aspects

Mobility within NTN, as well as to and from terrestrial networks, follows the same principles as in 5G terrestrial networks, with some additional considerations.

During mobility between NTN and Terrestrial Network, a UE is not required to connect to both NTN and Terrestrial Network at the same time. Also, a UE may support mobility between radio access technologies based on different orbits (geostationary or non-geostationary at different orbits).

Several features have been enhanced for both Idle/Inactive and Connected mobility to adapt existing mechanisms and procedures to the NTN systems.

6.8.1 Idle Mode and Inactive Mode Mobility

In both RRC Idle and RRC INACTIVE states, UE-controlled mobility based on network configuration is performed through cell selection and cell reselection procedures. Cell selection and reselection in NR is the baseline in NTN Idle/Inactive mode procedure for the three types of service links (Earth-fixed, quasi-Earth-fixed, and Earth-moving cells).

Cell selection is used to identify a suitable or acceptable cell for the UE to camp on. It is applicable after a UE is switched on, after a UE leaves RRC connected state, and after a UE returns to an area of coverage.

Cell selection is based on the following principles [16]:

- The UE NAS layer identifies a selected PLMN (and equivalent PLMNs, if any).
- The UE searches the supported frequency bands and for each carrier frequency, it searches and identifies the strongest cell. It reads cell broadcast information to identify PLMN(s) and other relevant parameters (e.g. related to cell restrictions).
- The UE seeks to identify a suitable cell; if it is not able to identify a "suitable" cell it seeks to identify an "acceptable" cell.
- A cell is "suitable" if: the measured cell attributes satisfy the cell selection criteria (based on DL radio signal strength/quality); the cell belongs to the selected/equivalent PLMN; cell is not restricted (e.g. cell is not barred/reserved or part of "forbidden" roaming areas).
- An "acceptable" cell is one for which the measured cell attributes satisfy the cell selection criteria and the cell is not barred.
- Among the identified suitable (or acceptable) cells, the UE selects the strongest cell, (then it "camps" on that cell).
- As signaled/configured by the radio network, certain frequencies could be prioritized for camping.

The cell selection criteria, i.e., Srxlev and Squal (known as "S" criteria) are specified in 3GPP TS 38.304.

Cell reselection is the mobility solution in RRC Idle and RRC INACTIVE modes. The principles of the procedure are the following:

- Cell reselection is always based on Cell Defining SSB (CD-SSBs) located on the synchronization raster.
- The UE makes measurements of attributes of the serving and neighbor cells to enable the reselection process. For the search and measurement of inter-frequency neighboring cells, only the carrier frequencies need to be indicated.
- Cell reselection identifies the cell that the UE should camp on. It is based on cell reselection criteria, which involves measurements of the serving and neighbor cells.
- Intra-frequency reselection is based on ranking of cells.
- Inter-frequency reselection is based on absolute priorities where a UE tries to camp on the highest priority frequency available.
- A neighbor cell list can be provided by the serving cell to handle specific cases for intra- and inter-frequency neighboring cells.
- Black lists can be provided to prevent the UE from reselecting to specific intra- and inter-frequency neighboring cells.
- Cell reselection can be speed-dependent.
- Service-specific prioritization.

In RRC Idle, a UE can autonomously perform cell reselection without informing the network, as long as it remains within a registered TA. The UE should acquire SIB1 after each cell reselection to determine whether or not the UE remains located within a registered TA. Upon changing to a new TA outside the UE's Registration Area, a mobility Registration Update procedure should be performed.

In Non-terrestrial networks, a TA corresponds to a fixed geographical area. Any respective mapping is configured in the RAN. The network may broadcast multiple TAC per PLMN in an NR NTN cell in order to reduce the signaling load at cell edge, in particular for Earth-moving cell coverage [4]. A TAC change in the System Information is under network control and may not be exactly synchronized with real-time illumination of beams on ground.

In 5G NR, specified rules [16] are used by the UE to limit needed measurements. These rules are illustrated in Figure 6.7:

For Intra-frequency Cell Reselection:
If the serving cell selection RX level value (Srxlev) and Cell selection quality value (Squal) are above some given thresholds; that is, Srxlev > **SIntraSearchP** and Squal > **SIntraSearchQ**, the UE may choose

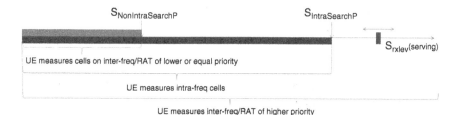

Figure 6.7   Measurement rules for cell reselection.

not to perform intra-frequency measurements. Otherwise, the UE shall perform intra-frequency measurements.

For Inter-frequencies and Inter-RAT Frequencies Cell Reselection:
The UE shall apply the following rules for NR inter-frequencies and inter-RAT frequencies, which are indicated in system information and for which the UE has priority:

- For an NR inter-frequency or inter-RAT frequency with a reselection priority higher than the reselection priority of the current NR frequency, the UE shall perform measurements of higher priority NR inter-frequency or inter-RAT frequencies according to TS 38.133.
- For an NR inter-frequency with an equal or lower reselection priority than the reselection priority of the current NR frequency and inter-RAT frequency with lower reselection priority than the reselection priority of the current NR frequency.

If the serving cell fulfills Srxlev > **SnonIntraSearchP** and Squal > **SnonIntraSearchQ**, the UE may choose not to perform measurements of NR inter-frequencies or inter-RAT frequency cells of equal or lower priority. Otherwise, the UE shall perform measurements of NR inter-frequencies or inter-RAT frequency cells of equal or lower priority according to TS 38.133.

In non-terrestrial networks, unlike terrestrial systems, there is a small difference in signal strength between two beams in a region of overlap. Further, Srxlev and Squal might not change too much while the serving cell remains active but drops immediately when the beam stops covering the area. The received signal strength in an NTN cell compared to a terrestrial network cell is shown in Figure 6.8. Relying on Srxlev or Squal thresholds to trigger measurements in NTN might not be sufficient. Therefore, time-based and location-based rules are introduced to limit needed cell reselection measurements. These new rules are used by the UE to decide when to start performing measurements as described in Sections 6.8.1.1 and 6.8.1.2.

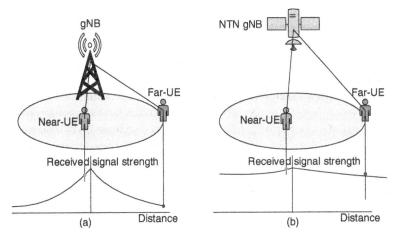

Figure 6.8   Received signal strength in (a) Terrestrial Network, (b) NTN.

6.8.1.1   Location-based Measurement

Location-assisted cell reselection, with the distance between UE and the reference location of the cell (serving cell and/or neighbor cell) taken into account, is supported in NTN cell. Thereby, to limit needed measurements, when the serving cell selection RX level value (Srxlev) and Cell selection quality value (Squal) are above-related thresholds, if the distance between UE and the serving cell reference location **referenceLocation** is shorter than **distanceThresh**, the UE may not perform measurements for cell reselection. These new rules assume that **distanceThresh** and **reference-Location** are broadcasted in SIB19, and the UE supports location-based measurement initiation [12] and has obtained its location information. When evaluating the distance between UE and the serving cell reference location, it's up to UE implementation to guarantee that a valid location information is available at the UE.

The parameters **distanceThresh** and **referenceLocation** are defined in Section 6.7.

6.8.1.2   Time-based Measurement

Time-based measurement is a new rule to be considered by the UE to decide when to start performing cell resection measurements within an NTN cell with quasi-Earth-fixed beams covering one geographic area for a limited period. It is based on the remaining time the cell is expected to remain active which is indicated in SIB19 with **t-Service**. This Release-17 parameter indicates the time information on when a cell provided via NTN quasi-Earth fixed system is going to stop serving the area it is currently covering. The parameter definition is given in Section 6.7.

As specified in TS 38.304, the UE shall perform intra-frequency, inter-frequency, or inter-RAT measurements before the t-Service, regardless of the distance between UE and the serving cell reference location or whether the serving cell selection RX level value (Srxlev) and Cell selection quality value (Squal) are above-related thresholds. UE shall perform measurements of higher priority NR inter-frequency or inter-RAT frequencies according to TS 38.133 regardless of the remaining service time of the serving cell (i.e., time remaining until *t-Service*). As specified in TS 38.306, it is optional for the UE to support time-based RRM measurements in RRC_IDLE/RRC_INACTIVE as specified in TS 38.304.

6.8.2 Connected Mode Mobility

In 5G NR, the term handover refers to the process of transferring an ongoing user connection from one radio channel to another. If active UE due to its movement can be served in a more efficient manner in another cell – a handover is performed. Network-controlled handover is applied to 5G NR UEs in RRC CONNECTED mode. This network-controlled handover is categorized into two types: cell level and beam level. Cell-level mobility requires explicit RRC signaling to be triggered, i.e., handover. For inter-cell handover, handover request, handover acknowledgment, handover command, and handover complete procedure are supported between source cell and target cell. The release of the resources at the source gNB during the handover completion phase is triggered by the target gNB.

Beam level mobility does not require explicit RRC signaling to be triggered – it is dealt with at lower layers – and RRC is not required to know which beam is being used at a given point in time.

The network may configure an RRC_CONNECTED UE to perform measurements and report them in accordance with the measurement configuration. The network may configure the UE to perform NR measurements and Inter-RAT measurements. It may configure the UE to report measurement information based on SS block(s) or based on Channel State Information Reference Signal (CSI-RS) resources (Table 6.2).

The criterion that triggers the UE to send a measurement report can either be periodical or a single event.

Measurement reports include the measurement identity of the associated measurement configuration that triggered the reporting. Cell and beam measurement quantities to be included in measurement reports are configured by the network. The number of non-serving cells to be reported

Table 6.2 Measurement events.

Measurement events:	
A1	Serving cell becomes better than threshold
A2	Serving cell becomes worse than threshold
A3	Neighbor becomes offset better than serving
A4	Neighbor becomes better than threshold
A5	Serving becomes worse than threshold1 and neighbor becomes better than threshold2
A6	Neighbor becomes offset better than secondary cell
B1	Inter-RAT neighbor becomes better than threshold
B2	PCell becomes worse than threshold1 and inter-RAT neighbor becomes better than threshold2
D1	Distance between UE and a reference location *referenceLocation1* becomes larger than configured threshold *distanceThreshFromReference1* and distance between UE and a reference location *referenceLocation2* becomes shorter than configured threshold *distanceThreshFromReference2*

can be limited through configuration by the network. Beam measurements to be included in measurement reports are configured by the network (beam identifier only, measurement result and beam identifier, or no beam reporting).

6.8.2.1 RRM Enhancements

RRM Event D1 RRM Event D1 is a new measurement triggering event based on location. It is introduced to trigger location reporting based on UE location in NTN. For this measurement event, the UE is configured with two reference locations and two threshold distances. The **referenceLocation1** and **distanceThreshFromReference1** can be associated with serving cell, and **referenceLocation2** and **distanceThreshFromReference2** can be associated with a neighbor cell.

The Event D1 is triggered when the distance between UE and referenceLocation1 is offset larger than distanceThreshFromReference1 and distance between UE and referenceLocation2 is offset shorter than distanceThreshFromReference2. As illustrated in Figure 6.9, UE shall consider the entering condition for this event to be satisfied when both $Ml1 - Hys >$ distanceThreshFromReference1 and $Ml2 + Hys <$ distance

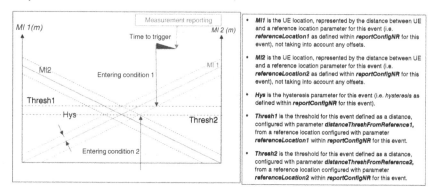

Figure 6.9 Entering condition of Event D1.

ThreshFromReference2, are fulfilled. And UE shall consider the leaving condition for this event to be satisfied when $Ml1 + Hys <$ distanceThresh FromReference1 or $Ml2 - Hys >$ distanceThreshFromReference2, i.e., at least one of the two, are fulfilled. Further, in order to trigger a measurement report-specific criteria for the event need to be met during a certain time duration defined by time to trigger parameter (i.e. **TimeToTrigger**).

SMTC Enhancement 3GPP Release-15 specification has introduced SSB-based Measurement Timing Configuration window (i.e. SMTC). When a Base Station configures the UE with measurements, it also configures the UE with an SMTC. The SMTC configuration provides the UE with the measurement periodicity and timings of SSBs that the UE can use for measurements during the window of the SMTC. The UE does not have to try to find the SSB outside of this window. The SMTC window periodicity can be set in the same range of SSB, i.e., 5, 10, 20, 40, 80, or 160 ms and window duration can be set to 1, 2, 3, 4, or 5 ms, according to the number of SSBs transmitted on the cell being measured.

In NTN, the propagation delay difference may have an impact on RRM measurements as highlighted in [11]. If the SMTC measurement configuration does not consider the propagation delay difference, e.g., between the UE may miss the SSB/CSI-RS measurement window and will thus be unable to perform measurements on the configured reference signals [11]. To address this issue, SMTC enhancement was specified in Release-17: the network can configure multiple SMTCs in parallel per carrier and for a given set of cells depending on UE capabilities using propagation delay difference and ephemeris information. It can also configure measurement gaps based on multiple SMTC. The adjustment of SMTCs is possible under network control based on UE assistance information if available for connected mode.

6.8.2.2 Conditional Handover

In general, a Conditional Handover (CHO) is defined as a handover, which is executed by the UE when one or more handover conditions are met. These conditions are part of the CHO configuration sent by the gNB to the UE [4].

The CHO configuration contains the configuration of candidate cells generated by the candidate target gNBs, and the execution conditions generated by the source gNB. For terrestrial networks, an execution condition may consist of up to two trigger conditions; at most, two different trigger conditions: A3 and A5 measurement-based events. If two CHO trigger conditions are configured, the UE may only trigger the handover if both CHO trigger conditions are fulfilled for one of the target candidate cells [4].

In a terrestrial network where the cells are characterized by a clear difference in signal strength between the cell center and the cell edge as illustrated in Figure 6.7b, it is expected that A3 and A5 measurement-based events are enough for an efficient CHO. However, in NTN triggering events based only on signal strength might not be efficient, e.g., at the border of a large NTN cell where the signal strength alone is not enough to distinguish which cell UE should be served from. Further, in NTN, a potentially very large number of UEs may need to perform HO at a given time, leading to possibly large signaling overhead and service continuity challenges, e.g., at service or feeder link switchover. Thereby, a time-based CHO trigger condition would allow to execute the handovers smoothly during longer time duration to avoid a massive number of HOs at the same time. This will also give information to the UEs on when the switch will occur.

Therefore, in NTN, new triggering conditions upon which UE may execute CHO to a candidate cell, have been introduced: A *time-based* condition and a *location-based* condition. As well as A4 event (Neighbor becomes better than threshold)-based trigger. Location is defined by the distance between UE and a reference location. Time is defined by the time between T1 and T2, where T1 is an absolute time value and T2 is a duration started at T1 [4].

The Base Station can still configure up to two trigger conditions per candidate target cell and one of these trigger conditions needs to be signal strength/quality measurement-related event.

6.9 Feeder Link Switchover

A feeder link switchover is the procedure where the feeder link is changed from a source NTN gateway to a target NTN gateway for a specific NTN

Figure 6.10 "Hard" feeder link switchover for non-geostationary satellites. Source: 3GPP TR 38.821 [11]/The TR is from 3GPP.

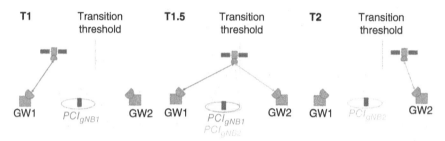

Figure 6.11 "Soft" feeder link switchover for non-geostationary satellites. Source: 3GPP TR 38.821 [11]/idem.

payload. The feeder link switchover is a TNL procedure; that is, it does not involve NG-RAN functions specified by 3GPP [4].

Both hard and soft feeder link switchovers are applicable to NTN. A hard switchover has only one feeder link serving the satellite (see Figure 6.10); a soft switchover has more than one feeder link serving the satellite at the same time (see Figure 6.11). Both hard and soft feeder link switchovers are applicable to NTN.

A feeder link switchover may result in transferring the established connection for the affected UEs between two gNBs [17].

For soft switchover, an NTN payload connects to more than one NTN gateway during a given period; that is, a temporary overlap can be ensured during the transition between the feeder links. For hard feeder link switchover, an NTN payload only connects to one NTN gateway at any given time; this implies that a radio link interruption may occur during the transition between the feeder links.

The NTN control function (see Section 6.6) determines the point in time when the feeder link switchover between two gNBs is performed. The transfer of the affected UE contexts between the two gNBs is performed through a handover; it depends on gNB implementation and configuration information provided to the gNBs by the NTN Control function [4].

6.10 Network Management Aspects

A more general representation of an NTN-based NG-RAN is shown in Figure 6.12. It shows some components not part of the 3GPP specification, including the NTN control function, the OAM, and other infrastructure components. These may be part of a typical NTN implementation [4].

The gNB in Figure 6.12 may be subdivided into non-NTN infrastructure gNB functions and NTN Service Link provisioning System. The NTN infrastructure consists of the *NTN Service Link provisioning System* and the *NTN Control function*. The NTN Service Link provisioning System may consist of one or more NTN payloads and NTN gateways.

The *NTN Service Link provisioning System* maps the NR-Uu radio protocol over radio resources provided by the NTN infrastructure (e.g., beams, channels, and TX power).

The *NTN control function* controls the spaceborne (or airborne) vehicles as well as the radio resources of the NTN infrastructure (NTN payload(s) and NTN gateway(s)). It provides control data, including ephemeris, to the non-NTN infrastructure gNB functions of the gNB.

At least the following NTN-related parameters are expected to be provided by OAM to the gNB for operation [4]:

(a) Earth fixed beams: For each beam provided by a given NTN payload
   o The Cell identifier (NG and Uu) mapped to the beam
   o The Cell's reference location (e.g. cell's center and range)
(b) Quasi-Earth fixed beams: For each beam provided by a given NTN payload:
   o The Cell identifier (NG and Uu) and time window mapped to a beam
   o The Cell's/beam's reference location (e.g. cell center and range)

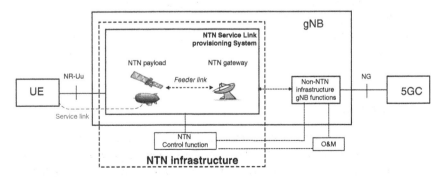

Figure 6.12 NTN-based NG-RAN [4].

    o The time window of the successive switchovers (feeder link and service link)

    o The identifier and time window of all serving satellites and NTN gateways

(c) Earth moving beams: For each beam provided by a given NTN payload:

    o The Uu Cell identifier mapped to a beam and mapping information to fixed geographical areas reported on NG, including information about the beam direction and motion of the beam footprint on Earth

    o Its elevation with respect to the NTN payload

    o Schedule of successive serving NTN gateways/gNBs

    o Schedule of successive switchovers (feeder link and service link)

In particular, the following parameters shall be provided by OAM to the gNB [4]:

- Ephemeris information describing the orbital trajectory information or coordinates for the NTN vehicles. This information is provided on a regular basis or upon demand to the gNB.
- Two different sets of ephemeris formats shall be supported:
  - o Set 1: Satellite position and velocity state vectors:
    - ■ Position
    - ■ Velocity
  - o Set 2: At least the following parameters in orbital parameter ephemeris format, as specified by [18]:
    - ■ Semimajor axis
    - ■ Eccentricity
    - ■ Argument of periapsis
    - ■ Longitude of ascending node
    - ■ Inclination
    - ■ Mean anomaly at epoch time t_0
- The explicit epoch time associated with ephemeris data
- The location of the NTN gateways
- Additional information to enable gNB operation for feeder/service link switchovers

A. Annex

Here is the complete mapping of standardized 5QI values to the various QoS characteristics as defined by 3GPP [2]. This includes provisions made for propagation delays, which may be encountered with LEO and GEO satellite systems (see footnote *q* of Table 6.A.1).

Table 6.A.1 Standardized mapping of 5QI to QoS characteristics.

5QI value	Resource type	Default priority level	Packet delay budget[a]	Packet error rate	Default maximum data burst volume[b]	Default averaging window	Example services
1	GBR	20	100 ms[c),d]	10^{-2}	N/A	2000 ms	Conversational voice
2	e)	40	150 ms[c),d]	10^{-3}	N/A	2000 ms	Conversational video (live streaming)
3		30	50 ms[c),d]	10^{-3}	N/A	2000 ms	Real-Time Gaming, V2X messages (see TS 23.287). Electricity distribution – medium voltage, process automation monitoring
4		50	300 ms[c),d]	10^{-6}	N/A	2000 ms	Non-conversational video (buffered streaming)
65[f),g]		7	75 ms[h),i]	10^{-2}	N/A	2000 ms	Mission Critical user plane Push To Talk voice (e.g. MCPTT)
66[g]		20	100 ms[d),j]	10^{-2}	N/A	2000 ms	Non-Mission-Critical user plane Push To Talk voice
67[g]		15	100 ms[d),j]	10^{-3}	N/A	2000 ms	Mission Critical Video user plane
75[k]							
71		56	150 ms[c),d),l]	10^{-6}	N/A	2000 ms	"Live" Uplink Streaming (e.g. TS 26.238)
72		56	300 ms[c),d),l]	10^{-4}	N/A	2000 ms	"Live" Uplink Streaming (e.g. TS 26.238)
73		56	300 ms[c),d),l]	10^{-8}	N/A	2000 ms	"Live" Uplink Streaming (e.g. TS 26.238)
74		56	500 ms[c),l]	10^{-8}	N/A	2000 ms	"Live" Uplink Streaming (e.g. TS 26.238)
76		56	500 ms[c),d),l]	10^{-4}	N/A	2000 ms	"Live" Uplink Streaming (e.g. TS 26.238)
5	Non-GBR	10	100 ms[d),j]	10^{-6}	N/A	N/A	IMS Signaling

(Continued)

Table 6.A.1 (Continued)

5QI value	Resource type	Default priority level	Packet delay budget[a]	Packet error rate	Default maximum data burst volume[b]	Default averaging window	Example services
6	[e]	60	300 ms[d),j]	10^{-6}	N/A	N/A	Video (Buffered Streaming) TCP-based (e.g. www, e-mail, chat, ftp, p2p file sharing, and progressive video)
7		70	100 ms[d),j]	10^{-3}	N/A	N/A	Voice, Video (Live Streaming) Interactive Gaming
8		80	300 ms[d]	10^{-6}	N/A	N/A	Video (Buffered Streaming) TCP-based (e.g. www, e-mail, chat, ftp, p2p file sharing, and progressive video)
9		90					
10		90	1100 ms[d),m]	10^{-6}	N/A	N/A	Video (Buffered Streaming) TCP-based (e.g. www, e-mail, chat, ftp, p2p file sharing, progressive video, etc.) and any service that can be used over satellite access type with these characteristics
69[f),g]		5	60 ms[h),i]	10^{-6}	N/A	N/A	Mission Critical delay sensitive signaling (e.g. MC-PTT signaling)
70[g]		55	200 ms[h),j]	10^{-6}	N/A	N/A	Mission Critical Data (e.g. example services are the same as 5QI 6/8/9)
79		65	50 ms[d),j]	10^{-2}	N/A	N/A	V2X messages (see TS 23.287)
80		68	10 ms[j),n]	10^{-6}	N/A	N/A	Low Latency eMBB applications Augmented Reality
82	Delay-critical GBR	19	10 ms[o]	10^{-4}	255 bytes	2000 ms	Discrete Automation (see TS 22.261)

5QI	Packet Delay Budget	Packet Error Rate	Maximum Data Burst Volume	Default Averaging Window	Example Services
83	10 ms[o)]	10^{-4}	1354 bytes[a)]	2000 ms	Discrete Automation (see TS 22.261 [2]);
22					V2X messages (UE – RSU Platooning, Advanced Driving; Cooperative Lane Change with low LoA. See TS 22.186, TS 23.287)
84	30 ms[p)]	10^{-5}	1354 bytes[a)]	2000 ms	Intelligent transport systems (see TS 22.261)
85	5 ms[n)]	10^{-5}	255 bytes	2000 ms	Electricity Distribution – high voltage (see TS 22.261). V2X messages (Remote Driving. See TS 22.186, footnote[q)], see TS 23.287)
86	5 ms[n)]	10^{-4}	1354 bytes	2000 ms	V2X messages (Advanced Driving; Collision Avoidance, Platooning with high LoA. See TS 22.186, TS 23.287)
87	5 ms[o)]	10^{-3}	500 bytes	2000 ms	Interactive Service – Motion tracking data (see TS 22.261)
88	10 ms[o)]	10^{-3}	1125 bytes	2000 ms	Interactive Service – Motion tracking data (see TS 22.261)
89	15 ms[o)]	10^{-4}	17 000 bytes	2000 ms	Visual content for cloud/edge/split rendering (see TS 22.261)
90	20 ms[o)]	10^{-4}	63 000 bytes	2000 ms	Visual content for cloud/edge/split rendering (see TS 22.261)

NOTE: It is preferred that a value less than 64 is allocated for any new standardized 5QI of Non-GBR resource type. This is to allow for option 1 to be used as described in clause 5.7.1.3 (as the QFI is limited to less than 64).

a) The Maximum Transfer Unit (MTU) size considerations in clause 9.3 and Annex C of TS 23.060 are also applicable. IP fragmentation may have impacts on CN PDB, and details are provided in clause 5.6.10.

b) It is required that default MDBV is supported by a PLMN supporting the related 5QIs.

c) In RRC Idle mode, the PDB requirement for these 5QIs can be relaxed for the first packet(s) in a downlink data or signaling burst in order to permit battery saving (DRX) techniques.

d) A static value for the CN PDB of 20 ms for the delay between a UPF terminating N6 and a 5G-AN should be subtracted from a given PDB to derive the packet delay budget that applies to the radio interface.

e) A packet that is delayed more than PDB is not counted as lost and, thus not included in the PER.

f) It is expected that 5QI-65 and 5QI-69 are used together to provide Mission Critical Push to Talk service (e.g. 5QI-5 is not used for signaling). It is expected that the amount of traffic per UE will be similar or less compared to the IMS signaling.

g) This 5QI value can only be assigned upon request from the network side. The UE and any application running on the UE are not allowed to request this 5QI value.

h) For Mission Critical services, it may be assumed that the UPF terminating N6 is located "close" to the 5G_AN (roughly 10 ms) and is not normally used in a long-distance, home-routed roaming situation. Hence a static value for the CN PDB of 10 ms for the delay between a UPF terminating N6 and a 5G_AN should be subtracted from this PDB to derive the packet delay budget that applies to the radio interface.

i) In both RRC Idle and RRC Connected mode, the PDB requirement for these 5QIs can be relaxed (but not to a value greater than 320 ms) for the first packet(s) in a downlink data or signaling burst in order to permit reasonable battery saving (DRX) techniques.

j) In both RRC Idle and RRC Connected mode, the PDB requirement for these 5QIs can be relaxed for the first packet(s) in a downlink data or signaling burst in order to permit battery saving (DRX) techniques.

k) This 5QI is not supported in this Release of the specification as it is only used for transmission of V2X messages over MBMS bearers as defined in TS 23.285 but the value is reserved for future use.

l) For "live" uplink streaming (see TS 26.238), guidelines for PDB values of the different 5QIs correspond to the latency configurations defined in TR 26.939. In order to support higher latency reliable streaming services (above 500 ms PDB), if different PDB and PER combinations are needed these configurations will have to use non-standardized 5QIs.

m) The worst case one way propagation delay for GEO satellite is expected to be ~270 ms, ~21 ms for LEO at 1200 km, and 13 ms for LEO at 600 km. The UL scheduling delay that needs to be added is also typically two way propagation delay e.g. ~540 ms for GEO, ~42 ms for LEO at 1200 km, and ~26 ms for LEO at 600 km. Based on that, the 5G-AN Packet delay budget is not applicable for 5QIs that require 5G-AN PDB lower than the sum of these values when the specific types of satellite access are used (see TS 38.300). 5QI-10 can accommodate the worst-case PDB for GEO satellite type.

n) A static value for the CN PDB of 2 ms for the delay between a UPF terminating N6 and a 5G-AN should be subtracted from a given PDB to derive the packet delay budget that applies to the radio interface. When a dynamic CN PDB is used, see clause 5.7.3.4.

o) A static value for the CN PDB of 1 ms for the delay between a UPF terminating N6 and a 5G-AN should be subtracted from a given PDB to derive the packet delay budget that applies to the radio interface. When a dynamic CN PDB is used, see clause 5.7.3.4.

p) A static value for the CN PDB of 5 ms for the delay between a UPF terminating N6 and a 5G-AN should be subtracted from a given PDB to derive the packet delay budget that applies to the radio interface. When a dynamic CN PDB is used, see clause 5.7.3.4.

q) These services are expected to need much larger MDBV values to be signaled to the RAN. Support for such larger MDBV values with low latency and high reliability is likely to require a suitable RAN configuration, for which, the simulation scenarios in TR 38.824 [3] may contain some guidance.

Source: Adapted from [2].

References

1 M. Olsson, S. Sultana, S. Rommer, L. Frid, C. Mulligan, *SAE and the Evolved Packet Core – Driving the Mobile Broadband Revolution*, Academic Press, 2009.

2 3GPP TS 23.501: "System Architecture for the 5G System (5GS); Stage 2", v. 17.5.0.

3 3GPP TR 38.824: "Study on physical layer enhancements for NR ultra-reliable and low latency case (URLLC)", v. 16.0.0.

4 3GPP TS 38.300: "NR; Overall description; Stage-2", v. 17.1.0.

5 3GPP TS 38.413: "NG-RAN; NG Application Protocol (NGAP)", v. 17.1.1.

6 3GPP TS 23.502: "Procedures for the 5G System (5GS); Stage 2", v. 17.5.0.

7 3GPP TS 38.410: "NG-RAN; NG general aspects and principles", v. 17.1.0.

8 3GPP TR 38.882: "Study on Requirements and User Cases for Network Verified UE Location for Non-Terrestrial-Networks (NTN) in NR", v. 18.0.0.

9 3GPP TS 38.305: "NG Radio Access Network (NG-RAN); Stage 2 functional specification of User Equipment (UE) positioning in NG-RAN", v.17.1.0.

10 3GPP TS 38.331: "NR; Radio Resource Control (RRC); Protocol specification", v.17.5.0.

11 3GPP TR 38.821: "Solutions for NR to support Non-Terrestrial Networks (NTN)", v. 16.1.0.

12 3GPP TS 38.306: "NR; User Equipment (UE) radio access capabilities", v.15.21.0.

13 3GPP TS 38.321: "NR; Medium Access Control (MAC) protocol specification," v.15.13.0.

14 H.-L. Määttänen, J. Sedin, S. Parolari, R. S. Karlsson, "Radio interface protocols and radio resource management procedures for 5G new radio non-terrestrial networks", *International Journal of Satellite Communications and Networking*, Special Issue, vol. 41, no. 3, pp. 276–288, May–June 2023.

15 3GPP TS 38.323: "Packet Data Convergence Protocol (PDCP) specification", v.17.5.0.

16 3GPP TS 38.304: "User Equipment (UE) procedures in Idle mode and RRC Inactive state", v.17.4.0.

17 G. Masini et al., "5G meets satellite: non-terrestrial network architecture and 3GPP", *International Journal of Satellite Communications and Networking*, Special Issue, pp. 1–13, 2022. doi: 10.1002/sat.1456.

18 NIMA TR 8350.2, 3rd Ed., Amendment 1: "Department of Defense World Geodetic System 1984", January 3, 2000. Available at: https://gis-lab.info/docs/nima-tr8350.2-wgs84fin.pdf.

7

RF and RRM Requirements

This chapter addresses radio frequency (RF) and radio resource management (RRM) aspects of Non-Terrestrial Networks (NTN) [1]. Among the topics described in this chapter, the reader can find information about the frequency bands in which NTN can operate, the NTN interfaces and architecture to which the RF requirements apply, the methodology used to define these RF requirements, and specific RRM requirements for the support of NTN.

7.1 Frequency Bands In Which NTN Can Operate

The 3rd Generation Partnership Project (3GPP) has defined as part of its Release-17, a number of enhancing features enabling 5G New Radio (5G NR), Narrowband Internet of things (NB-IoT), and enhanced Machine Type Communication (eMTC) radio interfaces to support NTN.

While these radio protocols are band agnostics, 3GPP defines the RF and the RRM requirements associated with a specific frequency band. A question is often raised about the frequency bands in which NTN can operate. In order to respond to this question, one should distinguish between satellite- and High-Altitude Platform System (HAPS)-based networks since the applicable regulations and related frequency bands are not the same.

7.1.1 Satellite Networks

Satellite networks, using the 3GPP-defined NTN radio interfaces on the service link, can operate in fixed or mobile satellite services (MSS) allocated bands [2].

5G Non-Terrestrial Networks: Technologies, Standards, and System Design, First Edition.
Alessandro Vanelli-Coralli, Nicolas Chuberre, Gino Masini, Alessandro Guidotti, and Mohamed El Jaafari.
© 2024 The Institute of Electrical and Electronics Engineers, Inc. Published 2024 by John Wiley & Sons, Inc.

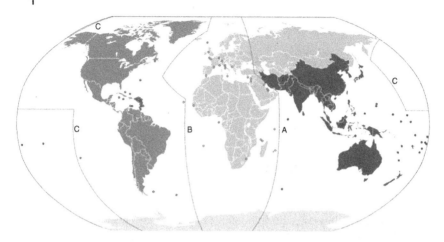

Figure 7.1 ITU regions and the dividing lines between them: Region 1 (Europe, Russia and africa), Region 2 (Americas), Region 3 (Asia and Oceania).

The satellite service allocated frequency bands in the different International Telecommunications Union (ITU) regions are depicted in Figure 7.1 and Table 7.1.

As part of the Release-17 and release independent work items, the bands in Table 7.2 have been defined for satellite networks using the NTN radio interface, while Table 7.3 provides those being defined in Release 18.

In addition, NTN could theoretically be deployed in mobile service allocated bands which is required to protect the operation of all other radio systems as per article 4.4 of the ITU-R Radio Regulations (RR). In line with this, the Federal Communications Commission (FCC, the US domestic regulator) announced on February 23, 2023 a future rulemaking process [3] by which "Supplemental Coverage from Space" could be provided in cellular bands by Non-Geosynchronous Orbit (NGSO) satellite operators on a basis of spectrum lease agreement with Mobile Network Operators (MNOs) having an exclusive terrestrial mobile license over certain geographical areas under US jurisdiction. The proposed rulemaking requires comments on all aspects: technical, regulatory, and competition. This is a significant move, as the FCC future ruling, if adopted, could pave the way for other administrations to pursue similar routes [4].

Table 7.1 Possible satellite service allocated frequency bands for NTN.

Bands	Direction	Region 1	Region 2	Region 3
« S-band » for GSO and Non-GSO space segment	Downlink (space to earth)	2170–2200 MHz	2160–2200 MHz	2170–2200 MHz
	Uplink (earth to space)	1980–2010 MHz	1980–2025 MHz	1980–2010 MHz
« C-band » for GSO and Non-GSO space segment	Downlink (space to earth)		3400–4200 and 4500–4800 MHz	
	Uplink (earth to space)	5725–7075 MHz	5850–7075 MHz	
« Ku-band » for GSO and Non-GSO space segment	Downlink (space to earth)	10.7–12.75 GHz	10.7–12.7 GHz	10.7–12.75 GHz
	Uplink (earth to space)		12.75–13.25 and 13.75–14.5 GHz	
« Ka-band » for GSO space segment	Downlink (space to earth)	17.3–20.2 GHz	17.7–20.2 GHz	
	Uplink (earth to space)	27.5–30.0 GHz	27.0–30.0 GHz	
« Ka-band » for Non-GSO space segment	Downlink (space to earth)	17.3–20.2 GHz	17.7–20.2 GHz	
	Uplink (earth to space)	27.5–29.1 and 29.5–30.0 GHz	27.0–29.1 and 29.5–30.0 GHz	
« Q/V-band » for GSO and Non-GSO space segment	Downlink (space to earth)	37.5–42.5, 47.5–47.9, 48.2–48.54, and 49.44–50.2 GHz	37.5–42.5 GHz	
	Uplink (earth to space)	42.5–43.5, 47.2–50.2, and 50.4–51.4 GHz		

When using the same bands allocated to the Fixed Satellite Service (FSS) or Broadcasting Satellite Service (BSS) for Geosynchronous Orbit (GSO) systems (e.g. C, Ku, KA, or Q/V bands), Non-GSO systems shall ensure protection of GSO systems.
The identified S bands in the table above can also be used by Complementary Ground Component (CGC) or Ancillary Terrestrial Component (ATC) of the network.
The band 17.3–17.7 GHz (space to earth) allocated in Region 1 may become also available in Regions 2 and 3, depending on future ITU WRC decisions.
The use of bands 47.5–47.9 GHz, 48.2–48.54 GHz, and 49.44–50.2 GHz in the downlink direction is limited to GSO systems.

Table 7.2 NTN operating bands in FR1 for satellite networks.

NTN satellite operating band	UpLink (UL) *operating band* SAN receive/UE transmit $F_{UL,low}$–$F_{UL,high}$	DownLink (DL) *operating band* SAN transmit/UE receive $F_{DL,low}$–$F_{DL,high}$	Duplex mode
n256	1980–2010 MHz	2170–2200 MHz	FDD
n255	1626.5–1660.5 MHz	1525–1559 MHz	FDD
n254	1610–1626.5 MHz	2483.5–2500 MHz	FDD

Table introduced in Chapter 1 and reported here for the sake of clarity. NTN satellite bands are numbered in descending order from n256 [5].

Table 7.3 NTN operating bands above 10 GHz for satellite networks.

NTN satellite operating band	UpLink (UL) *operating band* SAN receive/UE transmit $F_{UL,low}$–$F_{UL,high}$	DownLink (DL) *operating band* SAN transmit/UE receive $F_{DL,low}$–$F_{DL,high}$	Duplex mode
n512[a]	27.5–30.0 GHz	17.3–20.2 GHz	FDD
n511[b]	28.35–30.0 GHz	17.3–20.2 GHz	FDD
n510[c]	27.5–28.35 GHz	17.3–20.2 GHz	FDD

NOTE: Table introduced in Chapter 1 and reported here for the sake of clarity. The Downlink (DL) lower frequency range was modified from 17.7 to 17.3 GHz to reflect regional regulations. The notes were then treated individually:
a) This band is applicable in the countries subject to Conférence européenne des administrations des postes et télécommunications (CEPT) Electronic Communications Committee (ECC) Decision (05)01 and ECC Decision (13)01.
b) This band is applicable in the USA subject to FCC 47 CFR part 25.
c) This band is applicable for Earth Station operations in the USA subject to FCC 47 CFR part 25. FCC rules currently do not include Earth stations in motion (ESIM) operations in this band (47 CFR 25.202).

7.1.2 HAPS-based Networks

HAPS-based networks using the 3GPP-defined NTN radio interfaces on the service link can typically operate in mobile service-allocated bands according to ITU-R RR 5.388A and Resolution 221. For example, the bands in Table 7.4 as identified by ITU may be used by HAPS as base stations to provide International Mobile Telecommunications (IMT).

As part of the Release-17, the band in Table 7.5 has been defined for HAPS-based networks using NTN radio interfaces.

Table 7.4 Possible frequency bands for HAPS networks.

Duplexing mode	Direction	Region 1	Region 2	Region 3
FDD	Downlink (aerial to earth)	2110–2170 MHz	2110–2160 MHz	2110–2170 MHz
	Uplink (earth to aerial)	1885–1980 and 2010–2025 MHz	1885–1980 MHz	1885–1980 and 2010–2025 MHz
TDD	Up and Downlink (aerial to earth and earth to aerial)	2110–2170 MHz	2110–2160 MHz	2110–2170 MHz

Table 7.5 NTN operating bands in FR1 for HAPS networks.

NTN satellite operating band	UpLink (UL) *operating band* SAN receive/UE transmit $F_{UL,low}-F_{UL,high}$	DownLink (DL) *operating band* SAN transmit/UE receive $F_{DL,low}-F_{DL,high}$	Duplex mode
n1	1920–1980 MHz	2110–2170 MHz	FDD

7.2 NTN Architecture and Interfaces

Another question is how the 5G system architecture can be mapped over a NTN and to which interfaces the RF requirements are applicable. 3GPP has defined a reference architecture for satellite networks [5], which is illustrated in Figure 7.2.

The Satellite Access Node (SAN) corresponds to the Radio Access Network (RAN) of a terrestrial mobile system. SAN interfaces with the 5G Core (5GC) network via the NG interface and serves the User Equipment (UE) via the Uu interface of the New Radio (NR-Uu) interface. The 5G NR satellite system architecture is composed of NTN infrastructure and non-NTN infrastructure with gNB functionalities. This architecture further considers:

- The NTN service link provisioning system which maps the NR-Uu radio protocol over radio resources of the NTN infrastructure (e.g., beams, channels, and Tx power);

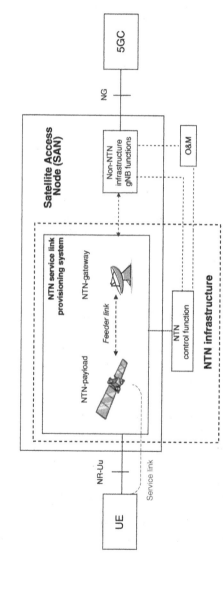

Figure 7.2 5G NR satellite system architecture in Release-17.

- The non-NTN infrastructure gNB functions, which implements the gNB logical function, possibly with different split options (e.g., gNB-DU/CU).

The NTN service link provisioning system is formed by (i) NTN payload(s) on board satellite(s), (ii) NTN Gateway(s) (NTN-GW), (iii) feeder link(s), which is a radio interface between the satellite and the NTN-GW. The NTN control function controls the radio resources of the SAN via Operation and Maintenance (O&M) and provides key information about the space segment such as the satellite Ephemeris information.

In particular, 3GPP defines the RF performance requirements applicable at the NR–Uu interface for both the SAN and the UE as physical entities.

In 3GPP Release-17 and 18, the satellite is assumed to embark a payload that transparently forwards the radio protocol received from the UE (via the service link) to the NTN Gateway (via the feeder link) and vice-versa. In future releases, the satellite may embark a regenerative payload implementing all or parts of the NG-RAN as well as some of the network functions of the 5GC.

7.3 Definition of RF Performances and Related Methodology

For all the bands introduced in 3GPP for NTN, the definition of the minimum RF characteristics and minimum transmission and reception performance requirements of both SAN and UE at the NR-Uu interface, shall not impact the existing 3GPP RF specifications applicable to bands already defined for NR and Long Term Evolution (LTE). In other words, the operation of NTN in these new bands shall not cause degradation to networks in 3GPP-specified terrestrial bands adjacent to the NTN band. Hence, this requires to undertake spectrum co-existence studies between NTN and terrestrial networks (TNs) operating in adjacent bands.

As described in [6], the method uses:

- Step 1: Identify relevant spectrum coexistence scenarios and analysis highlighting the most likely aggressor and victim between NTN and TNs for each of the candidate bands for NTN.
- Step 2: Discuss and agree on the most stringent case(s) for each spectrum coexistence scenario;
- Step 3: Discuss and determine the required Adjacent Channel Interference Ratio (ACIR) from results of the most stringent case(s) for each spectrum coexistence scenario;

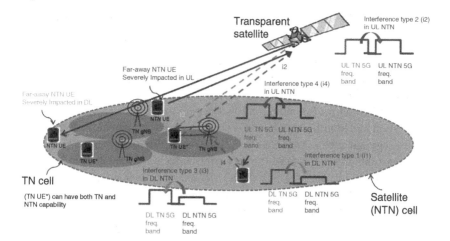

Figure 7.3 S-band NTN-TN adjacent band coexistence scenarios with TN in FDD mode (e.g. n1).

- Step 4: Use equation to derive corresponding ACLR or Adjacent Channel Selectivity (ACS) from the agreed ACLR for each spectrum coexistence scenario.

For example, the definition of the MSS allocated S-band or n256 led to consider the spectrum coexistence of NTN with a TN operating in adjacent frequency bands n1 and n34.

Six interference cases in Figure 7.3 (n256 coexistence with n1) and Figure 7.4 (n256 coexistence with n34) are represented. The four

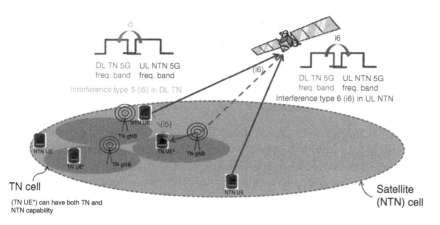

Figure 7.4 S-band NTN-TN adjacent band coexistence scenarios with TN in TDD mode (e.g. n34).

interference cases represented in Figure 7.3 (i.e., i1, i2, i3, and i4) and the two interference cases represented in Figure 7.4 (i.e., i5 and i6) are further described in detail in Table 7.7, with the explanations about the victim, the aggressor, and the interference direction (i.e., if UL or DL). As presented above, Figure 7.3 is applicable for satellites operating in S band, e.g., coexistence with TN n1 FDD, while Figure 7.4 is applicable for satellites operating in S band, e.g., coexistence with TN n34 TDD.

The main RF parameters directly obtained from coexistence analyses are based on Adjacent Channel Leakage Ratio (ACLR), which is a requirement at transmitter level, and ACS, which is a requirement at receiver level.

The main goal of the coexistence work in Release-17 was to define 5G NR NTN Satellite and UE RF requirements without any modification for the current 5G TN BS and UE requirements. The simulation considers two ACLR regions (ACLR1 and ACLR2) and one ACS value for the TN UE side, while for the TN BS only one value has been considered for ACLR and ACS, as explained in Table 7.6. For this reason, both TN gNB and UE ACLR/ACS parameters are kept static, while NTN Satellite and NTN UE ACLR/ACS parameters are varied. The purpose is to find the ACLR/ACS requirement values corresponding to less than 5% throughput loss, according to 3GPP methodology using ACIR as described in Table 7.7.

Table 7.7 provides a detailed explanation of the evaluation process and the simulation methodology. For instance, in the case of interference type i1 also represented in Figure 7.3, the TN gNB (the aggressor) may interfere in DL with the satellite NTN UE (the victim) receiving the same time information from the NTN satellite in DL, in adjacent band. Therefore, the ACIR to be evaluated is the ACIR of the NTN UE, which is a function of the TN gNB ACLR (transmitter, aggressor) and NTN UE ACS (receiver, victim). In order to evaluate this interference and the 5% throughput loss, the NTN UE ACS

Table 7.6 ACLR/ACS values for TN BS and TN UE (2 GHz).

Network entity	Requirement	New Radio (NR) values
Base Station (BS)	ACLR	45 dB
	ACS	46 dB
User Equipment (UE)	ACLR	30 dB (ACLR1); 43 dB (ACLR2)
	ACS	33 dB

Table 7.7 ACLR/ACS for TN (2 GHz).

Interference type	Combination	Aggressor	Victim	Evaluation	Simulation methodology
i1 (See Figure 7.3)	TN with NTN	TN DL	NTN DL	ACIR NTN UE with $\dfrac{1}{ACIR_{NTN\_UE}} = \dfrac{1}{ACLR_{TN\_gNB}} + \dfrac{1}{ACS_{NTN\_UE}}$	NTN UE ACS varied; TN gNB ACLR static
i2 (See Figure 7.3)	TN with NTN	TN UL	NTN UL	ACIR NTN Satellite with $\dfrac{1}{ACIR_{NTN\_Sat}} = \dfrac{1}{ACLR_{TN\_UE}} + \dfrac{1}{ACS_{NTN\_Sat}}$	NTN satellite ACS varied; TN UE ACLR static
i3 (See Figure 7.3)	TN with NTN	NTN DL	TN DL	ACIR TN UE with $\dfrac{1}{ACIR_{TN\_UE}} = \dfrac{1}{ACLR_{NTN\_Sat}} + \dfrac{1}{ACS_{TN\_UE}}$	NTN satellite ACLR varied; TN UE ACS static
i4 (See Figure 7.3)	TN with NTN	NTN UL	TN UL	ACIR TN gNB with $\dfrac{1}{ACIR_{TN\_gNB}} = \dfrac{1}{ACLR_{NTN\_UE}} + \dfrac{1}{ACS_{TN\_gNB}}$	NTN UE ACLR varied; TN gNB ACS static
i5 (See Figure 7.4)	TN with NTN	NTN UL	TN DL	ACIR TN UE with $\dfrac{1}{ACIR_{TN\_UE}} = \dfrac{1}{ACLR_{NTN\_UE}} + \dfrac{1}{ACS_{TN\_UE}}$	NTN UE ACLR varied; TN UE ACS static
i6 (See Figure 7.4)	TN with NTN	TN DL	NTN UL	ACIR NTN Satellite with $\dfrac{1}{ACIR_{NTN\_Sat}} = \dfrac{1}{ACLR_{TN\_gNB}} + \dfrac{1}{ACS_{NTN\_Sat}}$	NTN satellite ACS varied; TN gNB ACLR static

is varied while TN gNB ACLR is kept constant since one of the initial coexistence analyses was not to affect the existent 5G NR TN requirements in general (for both gNB and/or TN UE). The same reasoning can be used to provide explanations for the other scenarios in Table 7.7.

For each of those interference types identified in Table 7.7, 3GPP studied the coexistence between TN, equipped with Active Antenna System (AAS) or non-AAS antenna, and NTN, deployed at different altitudes as described in Table 7.11, in different propagation environments, i.e., rural or urban. From all of these combined scenarios, 3GPP selected to work on the scenarios represented in Table 7.8 in order to define the required ACIR for each interference type.

For the evaluation, the following radio interface configuration has been considered, with the channel bandwidths, Sub-Carrier Spacing (SCS), and Resource Block (RB) configurations represented in Table 7.9.

Table 7.8 Selected scenario for each interference type.

Scenario	Interference type	Aggressor system	Victim system	Environment	Contributing
1	i1	TN DL	NTN GEO DL	Urban	NTN UE ACS
2	i2	TN UL	NTN GEO UL	Urban	NTN SAN ACS
3	i3	NTN LEO-600 DL	TN DL	Rural	NTN SAN LEO ACLR
		NTN GEO DL	TN DL	Rural	NTN SAN GEO ACLR
4	i4	NTN GEO UL	TN UL	Urban	NTN UE ACLR
5	i5	NTN GEO UL	TN DL	Rural	NTN UE ACLR
6[a]	i6	NR-TN DL	NTN LEO-600 UL	Rural[b]	NTN SAN ACS
		NR-TN DL	NTN GEO UL	Rural[b]	NTN SAN ACS

a) Agreed representative case for Scenario 6.
b) The initial results suggested that the NR-NTN SAN would suffer more interference in urban deployment scenarios. It is agreed that a more relevant environment for case 6 is a mixture of Urban and Rural environments (e.g. urban area with a 50 km diameter inside a GEO beam with a 250 km diameter). Further studies based on the mixed urban environment could be considered. As compromise, rural only scenario was then selected.

Table 7.9 SAN channel bandwidths and SCS per operating band in FR1.

SAN operating band	SCS	SAN channel bandwidth and RB configuration			
		5 MHz	10 MHz	15 MHz	20 MHz
	15 kHz	25	52	79	106
n256	30 kHz	N/A	24	38	51
	60 kHz	N/A	11	18	24
	15 kHz	25	52	79	106
n255	30 kHz	N/A	24	38	51
	60 kHz	N/A	11	18	24
	15 kHz	25	52	79	N/A
n254	30 kHz	N/A	24	38	N/A
	60 kHz	N/A	11	18	N/A

7.3.1 Coexistence Analysis

Figure 7.5 present the worst-case coexistence study results for the victim system, as a function of applied ACIR values, averaged from all contributing sources for the selected scenario in Table 7.8. For example, the simulations considered the 5%-tile throughput loss or the average throughput loss, as explained in [6].

7.3.2 RF Performances

As presented in Table 7.10, Release-17 considers three satellite constellations for simulation and coexistence analysis: Geostationary Earth Orbit (GEO) satellite at 35 768 km altitude above Earth's Equator, Low Earth Orbit (LEO) satellite at 600 km altitude, and LEO satellite at 1200 km altitude.

The requirements have been further defined per SAN Class, as presented in Table 7.11. Therefore, based on the coexistence analysis results provided in the previous chapter, the agreed ACLR and ACS of NTN 5G NR Satellite and NTN UE are given in Table 7.11. One can notice that the minimum RF and performance requirements for NR UE operating in a NTN are identical to the ones of the NR UE operating in a TN, i.e., 30 dB of ACLR and 33 dB of ACS.

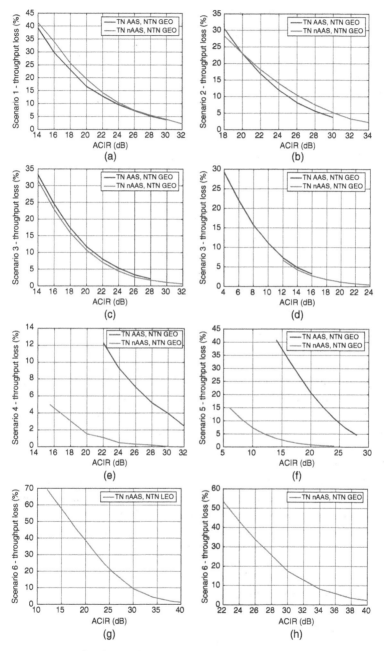

Figure 7.5 Throughput Loss (%) as a function of ACIR (dB) for TN-NTN coexistence analysis in adjacent bands: (a) scenario 1 (5%-tile throughput loss), (b) scenario 2 (average throughput loss), (c) scenario 3 with LEO (5%-tile throughput loss), (d) scenario 3 with GEO (5%-tile throughput loss), (e) scenario 4 (5%-tile throughput loss), (f) scenario 5 (5%-tile throughput loss), (g) scenario 6 with LEO (5%-tile throughput loss), and (h) scenario 6 with GEO (5%-tile throughput loss).

Table 7.10 SAN Classes in Release-17.

SAN class	Satellite constellation
GEO class	GEO satellite
LEO class	LEO 600 km satellite
	LEO 1200 km satellite

Table 7.11 ACLR and ACS of 5G NR NTN SAN and NTN UE in Release-17.

NTN 5G NR parameters			Values
SAN	ACLR	GEO Class	14 dB
		LEO Class	24 dB
	ACS[a]	GEO Class	38 dB
		LEO Class	38 dB
UE	ACLR		30 dB
	ACS		33 dB

a) The ACS values for SAN apply to both Rural and Urban environments.

The NTN-capable NR UE minimum RF and performance requirements are therefore aligned with the TN-capable NR UE minimum RF and performance requirements. For this reason, a UE implementing TS 38.101-5 [7] functionalities for connectivity with NTN can also implement TS 38.101-1 [8] functionalities for connectivity with the TN.

TS 38.101-1 [8] establishes the minimum RF requirements for "terrestrial" NR UE operating on frequency Range 1, while TS 38.101-4 [9] establishes the minimum performance requirements for "terrestrial" NR UE. In the case of NTN, TS 38.101-5 [7] specifies both minimum RF requirements and minimum performance requirements for NTN NR UE. The SAN specification TS 38.108 [5] which provides the SAN has the same structure as the TS 38.104 [10], which specifies the "terrestrial" gNB.

7.4 RRM Requirements

7.4.1 System Aspect

As discussed in Chapters 4, 5, and 6, 3GPP considered different potential methods for synchronization and timing advance. More precisely, in Release-17, RAN1 considered UE pre-compensation methods based on UE

Global Navigation Satellite System (GNSS) information, ephemeris data from the satellite, e.g., Position Velocity and Time (PVT) and/or orbital information, and a propagation delay information on the feeder link. For this reason, it is important to mention how the ephemeris data could be generated from predicted satellite position and velocity.

As introduced in Chapter 3, the satellite Orbit Determination (OD) consists in estimating the most likely track or orbit taken by a satellite, based on past and noisy measurements of its position and velocity. The satellite OD is useful for several reasons. It helps estimating what was the satellite's past trajectory so one can identify whether the satellite is diverging from a reference orbit, and if needed, schedule satellite station-keeping maneuvers to get it back on track. An overview of a typical satellite system architecture is given in Figure 7.6, where the locations of each OD step are explicitly represented. In general, the satellite OD is performed as follows:

Step 1: Measurements of the satellite position and velocity are made and dated. Most of the time, these measurements are GNSS-based measurements performed on-board. The accuracy of on-board satellite GNSS measurements depends on several factors such as:

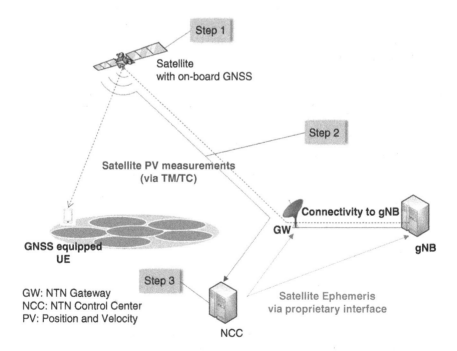

Figure 7.6 System architecture overview.

- The GNSS receiver characteristics (mono-frequency versus bi-frequency, single versus multi-constellation, quality of the GNSS receiver);
- The satellite orbit, and the number of GNSS satellites in view.

Step 2: The measurements are collected in the NTN Control Center (NCC), via TeleMetry/TeleCommand (TM/TC) functionalities. The satellite OD is further performed in the NTN Control Center. In general, the satellite GNSS measurements are delivered to the ground via dedicated telemetry channels between the satellite and the NTN GW network. The reporting period of new GNSS measurements is implementation-specific but it is directly related to the accuracy target associated with the instantaneous knowledge of the satellite position and velocity in the system.

Step 3: The satellite OD is performed in the NCC. This operation can be more or less complex depending on the models considered, the quantity of measurements available, and the algorithms used. Traditionally, two filtering techniques are used on input measurements: Kalman filtering or Least Square (LS) filtering. LS filtering usually provides more accurate results but requires more computing power.

The UE requirements on uplink transmission timing took into account the system architecture overview presented in Figure 7.6. Moreover, the total NTN timing error T_{e_NTN} can be expressed as a function of legacy timing error T_e, satellite timing error T_{e_Sat}, and GNSS error T_{e_GNSS}, as represented in the equation below:

$$T_{e\_NTN} = T_e + T_{e\_SAT} + T_{e\_GNSS}$$

Based on this approach, in Release-17 TS 38.133 [11], the timing error limit T_{e_NTN} requirement has been defined according to Table 7.12:

Table 7.12 T_{e_NTN} timing error limit.

Frequency range	SCS of SSB signals (kHz)	SCS of uplink signals (kHz)	T_{e_NTN}
1	15	15	$29*64*T_c$
		30	$24*64*T_c$
		60	N/A
	30	15	$24*64*T_c$
		30	$22*64*T_c$
		60	N/A

T_{e_NTN} is the timing error limit defined in TS 38.211 [12].

Table 7.13 T_{q_NTN} maximum autonomous time adjustment step and T_{p_NTN} minimum aggregate adjustment rate.

Frequency range	SCS of uplink signals (kHz)	T_{q_NTN}	T_{p_NTN}
1	15	$5.5*64*T_c$	$5.5*64*T_c$
	30	$5.5*64*T_c$	$5.5*64*T_c$
	60	N/A	N/A

NOTE: T_c is the basic timing unit defined in TS 38.211 [12].

Hence, the maximum autonomous time adjustment step T_{q_NTN} and minimum aggregate adjustment rate T_{p_NTN} are defined as represented in Table 7.13.

References

1 D. Panaitopol, Y. Jin, R. Tang, C. Park. "3GPP Advancements on 5G NR NTN Satellite Access Node and UE RF and RRM Aspects in Release-17".

2 RP-180135 Non Terrestrial Networks update on spectrum S band Geosynchronous Equatorial Orbit, Hughes Network Systems, Dish network, Thales (submitted at 3GPP TSG RAN Meeting #79, Chennai, India, 19–22 March 2018).

3 Single Network Future: "Supplemental Coverage from Space, Space Innovation, Notice of Proposed Rulemaking". https://www.fcc.gov/document/fcc-proposes-framework-facilitate-supplemental-coverage-space-0.

4 Horizon 2020 DYNASAT project D6.6 Standardization & Regulatory Report – Version 2. Available at: https://www.dynasat.eu/public-deliverables/.

5 3GPP TS 38.108: "Satellite Access Node radio transmission and reception (Release 17)".

6 3GPP TR 38.863: "Solutions for NR to support non-terrestrial networks (NTN): Non-terrestrial networks (NTN) related RF and co-existence aspects (Release 17)".

7 3GPP TS 38.101-5: "NR; User Equipment (UE) radio transmission and reception; Part 5: Satellite access Radio Frequency (RF) and performance requirements".

8 3GPP TS 3GPP 38.101-1: "NR; User Equipment (UE) radio transmission and reception; Part 1: Range 1".

9 3GPP TS 38.101-4: "NR; User Equipment (UE) radio transmission and reception; Part 4: Performance requirements".

10 3GPP TR 38.104: "NR; Base Station (BS) radio transmission and reception".

11 3GPP TS 38.133: "NR; Requirements for support of radio resource management".

12 3GPP TS 38.211: "NR; Physical channels and modulation".

8

NB-IoT and eMTC in NTN

8.1 Overview

Today, Internet of Things (IoT) communications are a pivotal element of worldwide society and economy. They allow objects to connect and communicate with each other, without depending on human intervention, but rather providing to humans the outcomes of their decisions, which can include alarms (e.g. monitoring IoT services) or sensor data (e.g. environmental monitoring). The vertical markets in which IoT services are providing a key added value are many-faceted, including transportation, logistics, smart agriculture, smart cities, smart buildings, connected healthcare, just to name a few [1]. In general, the IoT world can be classified into: (i) Massive IoT, characterized by a massive number (i.e. in the order of billions) of worldwide distributed devices that generate a small amount of traffic to be regularly transmitted to the core network, e.g., a set of sensors in a smart city or building; and (ii) Critical IoT, involving less terminals compared to Massive IoT but handling larger data packets, which provides Critical Communications services, e.g., remote monitoring of critical infrastructures or traffic monitoring, requiring ultra-high reliability, availability, and low latency. In general, despite the profound difference in the very nature of these services, the IoT requirements, and the related challenges, can be classified as follows:

- Low cost and low complexity: Limiting the complexity of the device and, thus, the cost of its hardware components are key enablers for mass-market, massive-volume applications. These can be achieved by design, e.g., by including a single Radio Frequency (RF) chain, reducing the supported bandwidth, and restricting the supported peak data rates. It shall be noticed that these aspects might be challenging for the provision of IoT services via Non-Terrestrial Network (NTN), in particular,

5G Non-Terrestrial Networks: Technologies, Standards, and System Design, First Edition.
Alessandro Vanelli-Coralli, Nicolas Chuberre, Gino Masini, Alessandro Guidotti, and Mohamed El Jaafari.
© 2024 The Institute of Electrical and Electronics Engineers, Inc. Published 2024 by John Wiley & Sons, Inc.

in terms of link budget, due to the low transmission power and low antenna gains at the IoT terminals, and synchronization and phase noise, due to the use of low-cost oscillators. Another feature that might pose challenges in terms of the terminal cost is related to the presence of a Global Navigation Satellite System (GNSS) receiver.

- Energy efficiency: The vast majority of IoT devices is battery-powered and they shall maintain their operability for a long period (i.e. years) without requiring human intervention. The battery life is clearly impacted by the power consumption during the transmission and reception phases and during the idle/sleep periods. To limit such battery drains, limited signaling during the idle/sleep should be considered, combined with a fast (re-)acquisition process when the terminal moves to the transmission or reception status. To this aim, when considering NTN implementation, ancillary information for the terminal (e.g. GNSS data and satellite ephemeris) might be beneficial.
- Support massive connectivity: depending on the coverage area, it can be expected that some cells are densely populated, i.e., hot-spots with thousands of devices per square km, compared to others. Thus, the system shall support the simultaneous handling of these devices. To this aim, it shall be mentioned that, given the sporadic burst nature of the IoT communications, fixed resource allocation plans might not be particularly efficient; as such, flexible payloads with dynamic resource allocation schemes shall be envisaged.
- Extreme coverage: This is a key requirement for massive IoT communications taking into account remote and scarcely populated areas. The exploitation of a space/aerial component in the overall infrastructure might be clearly beneficial, compared to purely terrestrial networks.
- Latency: For Critical IoT services, the minimization of the end-to-end latency is one of the key features. In this context, the exploitation of low altitude (e.g. LEO or HAPS) systems might be the best options. However, this poses additional challenges related to (i) the limited visibility period of each satellite, which is a function of the platform altitude and can also be as low as a few minutes; and (ii) the need for fast handover procedures, in particular when Earth moving beams are considered. Moreover, it shall be noticed that, depending on the selected Air Interface and system architecture (SA), assessing the impact of the over-the-air delay on the involved protocol procedures and timers is of paramount importance.
- Security and privacy: In general, the identity of the IoT terminal should not be publicly available, but still, it shall be traceable by the authorities. In this framework, it shall be mentioned that, from a cyber-security

perspective, the overall protection level of the network is defined by the element with the lowest security, which most likely is the IoT terminal, if present.

8.1.1 Cellular IoT in 3GPP Roadmap

The 3GPP standardization framework ensures a global ecosystem around Cellular Internet of Things (CIoT) and provides for proven security mechanisms. The IoT requires low-cost power-efficient global connectivity services. New physical layer (PHY) solutions, MAC procedures, and network architectures were considered in CIoT design, leveraging the legacy Long-Term Evolution (LTE) cellular systems to meet the demands of IoT services. Several steps have been taken under the 3GPP to accomplish these objectives. Different releases of LTE have provided progressively improved support for low-power wide-area IoT connectivity, specifically: Network Improvements for Machine Type Communications (MTC) in 3GPP Release-10 and System Improvements for MTC in Release-11. 3GPP Release-12 has specified low-cost M2M devices (Category 0).

Substantial Evolved Universal Terrestrial Radio Access Network (E-UTRAN)/Evolved Packet Core (EPC) evolution has been achieved in 3GPP to enable the CIoT. In particular, enhanced Machine Type Communication (eMTC) and Narrow Band IoT (NB-IoT) have been designed in RAN WGs in Release-13 and enhanced in Release-14. The corresponding SA aspects have been designed for EPC in Release-13 and Release-14. These SA aspects apply to both NB-IoT and eMTC. Thereby, two technologies have been introduced to support narrow-band MTC: eMTC and NB-IoT.

NB-IoT has been specified to provide low-cost, low-power, wide-area cellular connectivity for the IoT. Standardization of NB-IoT was completed in June 2016. NB-IoT is to a large extent a new radio access technology based on LTE technology. It supports most LTE functionalities albeit with essential simplifications to reduce device complexity. Further optimizations to increase coverage, reduce overhead, and reduce power consumption while increasing capacity have been introduced as well. The design objectives of NB-IoT include low complexity devices, high coverage, long-device battery life, and massive capacity. Latency is relaxed, although a delay budget of 10 seconds is the target for exception reports. NB-IoT can operate over a system bandwidth as low as 200 kHz, in stand-alone mode, within an LTE carrier (i.e. in-band operation, which means that operators can run NB-IoT on their existing LTE frequency allocations) or, within the guard-band of an LTE carrier. It also supports a minimum channel bandwidth of only 3.75 kHz. Two UE categories were defined for NB-IoT: Cat-NB1 in

Release-13 and Cat-NB2[1] in Release-14. This gives an unmatched spectrum flexibility and system capacity, which in combination with qualities such as energy efficient operation, ultra-low device complexity, and ubiquities coverage makes NB-IoT a very competitive technology in the IoT market.

eMTC (a.k.a LTE-M which stands Long Term Evolution for Machines) expands the possible applications by using six times the bandwidth of NB-IoT to enable data rates of up 1 Mbps and the integration of voice. Standardization of eMTC as part of Release-14 was completed in June 2017. eMTC provides improved both indoor and outdoor coverage, supports massive numbers of low throughput devices, low delay sensitivity, ultra-low device cost, low-device power consumption, and optimized network architecture. The main components of LTE-M are a series of low-cost device categories (e.g. Cat-M1[2] and Cat-M2[3]) and two coverage enhancement modes (i.e. CE modes A and B). eMTC was originally designed to reduce the device complexity to make LTE competitive with Enhanced General Packet Radio Service (EGPRS) in the MTC market. In addition to its low complexity, it does support secure communication, ubiquitous coverage, and high system capacity. eMTC's ability to operate as a full-duplex system over a larger bandwidth also gives it an additional dimension with its capability to offer services of lower latency and higher throughput than EC-GSM-IoT and NB-IoT, qualities which allow eMTC to support services such as voice over IP. eMTC supports both full-duplex frequency division duplex (FDD) and half-duplex FDD.

These 3GPP CIoT technologies were further enhanced in subsequent 3GPP Releases and became the global dominant technologies that are set to enable the huge market growth. Several enhancements and new features are introduced for NB-IoT in 3GPP LTE Release-15 and Release-16 to improve the user experience as well as cater to more use cases. Further NB-IoT enhancements in Release-15 is a collection of additions and enhancements of functionalities related to NB-IoT, primarily focusing on reducing UE power consumption and on enhancing the parts of the Rel-13/14 NB-IoT air interface and protocol layers to respond to feedback from early deployments. Additional new features include support for small cells, extensions to NB-IoT standalone operation mode, and Time Division Duplexing (TDD).

---

1 UE category Cat. NB2 is introduced in Release-14. It can support up to 2 HARQ processes and larger transport block size. This will allow Cat. NB2 UE to reach peak data rates of 127 kbps in the downlink and 159 kbps in the uplink.
2 In Release 14, Cat. M1 UE can support larger transport block sizes using the same 1.4 MHz bandwidth, which will increase peak rates to 3 Mbps in the uplink.
3 Cat. M2 supporting 5 MHz bandwidth is introduced in Rel-14. This UE category can support peak rates up to 4 Mbps in the downlink and 7 Mbps in the uplink.

The major enhancements are just to cite a few: Wake-up signaling for IDLE mode, Early data transmission, NB-IoT Physical Random Access Channel (NPRACH) range enhancement, and small cell support (FDD and TDD).

Even further enhanced MTC for LTE Release-15 work item (WI) builds on the LTE features for MTC introduced in Rel-13 and Rel-14 (e.g. low-complexity UE categories M1 and M2, and Coverage Enhancement Modes A and B) by adding support for new use cases and general improvements with respect to latency, power consumption, spectral efficiency, and access control. The new MTC features for LTE in Rel-15 include among other enhancements: Support for higher UE velocity (240 km/h at 1 GHz and 120 km/h at 2 GHz), lower UE power class (a new lower UE power class with a maximum transmission power of 14 dBm is introduced for Cat-M1 and Cat-M2), reduced UE power consumption and latency, increased spectral efficiency (e.g. downlink 64QAM support), and improved access control.

As part of Realese-16, the "Cellular IoT support and evolution for the 5G System" WI focused on enabling equivalent functionality for NB-IoT and eMTC connected to 5GC as what has been defined for NB-IoT and eMTC connected to EPC in earlier releases. Several CIoT features have been introduced in Release-16: Control Plane/User Plane (CP/UP) CIoT 5GS Optimization, non-IP Data Delivery, reliable Data Service (RDS), extended Discontinuous Reception (DRX) for CM-IDLE, and CM-CONNECTED with RRC-INACTIVE, high latency communication, support for monitoring events, system aspects of the enhanced coverage RAN feature, differentiation of Category M UEs, and selection, steering, and redirection between EPS and 5GS.

In Release-17, 3GPP conducted a WI on "additional enhancements for NB-IoT and LTE-MTC" to introduce features such as 16QAM for NB-IoT in downlink and uplink, 14 HARQ processes in downlink for HD-FDD Cat. M1 UEs, NB-IoT neighbor cell measurement and triggering before RLF, NB-IoT carrier selection based on coverage level, and a maximum DL TBS of 1736 bits for HD-FDD Cat. M1 UEs. Further, 3GPP carried out a study on NB-IoT/eMTC support for NTN, followed by a normative work on NB-IoT/eMTC support for NTN, which specifies enhanced features necessary for the support of Bandwidth reduced Low complexity (BL) UEs, UEs in enhanced coverage, and NB-IoT UEs by NTN. An overview on these study item (SI) and WI is provided in Sections 8.1.2 and 8.1.3. The reference deployments scenarios, and overall CIoT NTN architecture are described in Section 8.2. The necessary enhancements for NB-IoT/eMTC support in NTN specified in Release-17 are described in Section 8.3.

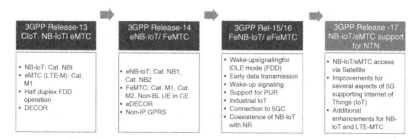

Figure 8.1 Cellular IoT in 3GPP Roadmap.

Figure 8.1 summarizes the major milestones of CIoT normative work in 3GPP Roadmap.

8.1.2 Study Item on IoT NTN

IoT NTN is defined in 3GPP as an E-UTRAN consisting of E-UTRAN Node B (eNB), which provide non-terrestrial LTE access to UEs by means of an NTN payload embarked on a space-borne NTN vehicle and an NTN gateway (GW). A SI to enable the provision of IoT services via NTN was approved during Radio Access Network (RAN) meeting #86 (and revised during RAN meeting #91 [2]) aiming at (i) identifying the scenarios that can be applied to NB-IoT and enhanced Machine Type Communications (eMTC); and (ii) based on these scenarios, identifying the recommended modifications on existing specifications to support NB-IoT/eMTC via NTN. With respect to the latter, the objective is to also maintain as much as possible the alignment with the studies reported in 3GPP TR 38.821 [3], and with the WI on Rel-17 NTN [4]. The following assumptions are considered for the analyses to be performed under the SI: (i) the User Equipment (UEs) have GNSS capabilities, allowing the terminal to estimate and, then, pre-compensate the time and frequency offset for uplink transmissions (the simultaneous operation of GNSS and NB-IoT/eMTC is not considered); (ii) the IoT features defined up to Rel.-16 shall be supported; and (iii) NB-IoT shall support both single- and multi-carrier operations. At system-level, the focus is on bands below 6 GHz and on transparent payloads, with Low Earth Orbit (LEO), Medium Earth Orbit (MEO) (lower priority), and GEO satellites. Moreover, it is assumed that no Inter-Satellite Links (ISL) are present.

The outcomes of the SI are reported in 3GPP TR 36.763 [5]. Similarly to the work related to NTN in TR 38.821, this technical report provides an overview of the scenarios, the architecture, and the challenges related to the procedures when the NTN radio link is considered, including the impact on PHY, MAC, Radio Link Control (RLC), and Packet Data Convergence

Protocol (PDCP) layers. In particular, TR 36.763 contains aspects related to Random Access (RA) procedure, general aspects related to RLC/MAC and PDCP timers, mobility, Radio Link Failure, Hybrid Automatic Repeat request (HARQ) operation, and time/frequency adjustments.

8.1.3 Normative Work on IoT NTN

As part of 3GPP Release-17, the "NB-IoT/eMTC support for Non-Terrestrial Networks (NTN)" WI specifies enhanced features necessary for the support of Bandwidth reduced Low complexity[4] (BL) UEs, UEs in enhanced coverage and NB-IoT UEs by NTN. The support of CIoT NTN in 3GPP Release-17 specifications is to a large extent aligned with that of NR-NTN in 5GS. The focus was on standalone deployment for NB-IoT/eMTC, i.e., operating in carrier(s) used only for NB-IoT NTN (resp. eMTC NTN) for support in Release-17 timeframe. All IoT NTN devices/UEs are supposed to be equipped with a GNSS receiver. With this assumption, UE can estimate and pre-compensate timing and frequency offset with sufficient accuracy for uplink transmission. Simultaneous GNSS and NTN NB-IoT/eMTC operation is not assumed. For less impacts on existing specifications, NB-IoT/eMTC design for terrestrial networks is reused as much as possible.

SA and Core network and Terminals (CT) aspects of NB-IoT/eMTC NTNs in Evolved Packet System (EPS) provide minimum essential functionality for the Release-17 UE and the network to support satellite E-UTRAN access in WB-S1[5] mode or NB-S1[6] mode with CIoT EPS optimization. The functionality is largely aligned with that of Release-17 NR NTN in 5GS, with the exception of discontinuous coverage that is addressed only within IoT NTN WI in Release-17.

8.2 Architecture and Deployments Scenarios

8.2.1 Potential Use Cases

The NB-IoT/eMTC-based satellite access aims at providing MTC services to IoT devices for applications in agriculture, transport, logistics, and security

---

4 BL UE can operate in any LTE system bandwidth but with a limited channel bandwidth of 6 PRBs (corresponding to the maximum channel bandwidth available in a 1.4 MHz LTE system) in DL and UL.
5 WB-S1 mode: The system operates in S1 mode, but not in NB-S1 mode.
6 NB-S1 mode: Serving radio access network provides access to network services via E-UTRA by NB-IoT.

markets. A preliminary classification for massive Machine Type Communications (mMTC) was provided when first identifying the 5G services that could be provided via satellite as part of the SI that was conducted by 3GPP in Release-15. The outcomes of this SI are collected in the TR 38.811 [6]:

• *Wide area IoT*, for the global continuity of service for applications, including groups of fixed or moving sensors and/or actuators deployed over a wide coverage area. These devices report the information to a central server, or are controlled by it, and they can be envisaged to be used for the following verticals:
  o Automotive and road transportation: High density platooning, High Definition (HD) map updates, traffic flow optimization, vehicle software updates, automotive diagnostic reporting, user base insurance information (e.g. speed limit and driving behavior), safety status reporting (e.g. air-bag deployment reporting), advertising-based revenue, context awareness information (e.g. neighboring bargain opportunities based on revenue), remote access functions (e.g. remote door unlocking).
  o Energy: Surveillance of critical infrastructures, e.g., oil/gas long distance pipes.
  o Transportation: Fleet management, asset tracking, and remote road alerts.
  o Smart agriculture: Livestock management, farming.
• *Local area IoT*, in which a group of sensors collect local information, connect to each other, and then report the data to a central entity. Such entity might also command a set of actuators, so as to take local actions. The sensors/actuators served by a Local Area Network (LAN) may be located in a smart grid sub-system (advanced metering) or on board a moving platform (e.g., container on board a vessel, a truck, or a train).

8.2.2 System Architecture

3GPP Release-17 specified the following architecture for NB-IoT/eMTC in NTN: The EUTRAN in Figure 8.2 consists of a set of Satellite Access nodes (SAN) connected to the EPC via the S1 interface. Only Bandwidth reduced Low complexity (BL) UEs are considered, with support of UEs in enhanced coverage and NB-IoT UEs with GNSS capabilities.

As illustrated in Figure 8.2, a SAN provides the EUTRAN User Plane (UP) and Control Plane (CP) terminations toward an NTN-enabled IoT device, which can access the NTN services through the NTN payload via the service link, and it includes: a transparent NTN payload on-board the NTN platform, a GW interconnected by a feeder link, and eNB functions.

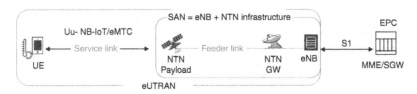

Figure 8.2 IoT NTN EUTRAN architecture.

As for NR NTN, three types of service links are supported for IoT NTN: Earth-fixed, Quasi-Earth-fixed, and Earth-moving. With NGSO satellites, the eNB can provide either quasi-Earth-fixed cell coverage or Earth-moving cell coverage, while eNB operating with GSO satellites can provide Earth-fixed cell coverage or quasi-Earth-fixed cell coverage.

Note that support for BL UEs, UEs in enhanced coverage and NB-IoT UEs over NTNs in Release-17 and Release-18 is only applicable to E-UTRA connected to EPC.

The UE Power Classes in Table 8.1 define the maximum output power for any transmission bandwidth within the channel bandwidth that are supported in IoT NTN for category M1, NB1, and NB2.

NTNs encompasses platforms that provide radio access through satellites in Geosynchronous orbits (GSO) as well as Non-Geosynchronous Orbit (NGSO), which includes LEO and MEO. The 3GPP's work on IoT NTN considers the scenarios reported in Table 8.2.

Apart from satellite parameters Set 1 shown in Table 8.3 (based on TR 38.821, Table 6.1.1.1-1) and Set 2 shown in Table 8.4 (based on TR 38.821, Table 6.1.1.1-2) originally used for NR NTN, two additional configuration sets have been included for LEO and GEO satellites to achieve extended coverage areas in IoT NTN. These two sets are referred to as Set-3 and Set-4 in [5] and reported in Tables 8.5 and 8.6, respectively.

Finally, a fifth configuration was proposed for MEO systems, and the related parameters are provided in Table 8.7.

Table 8.1 UE maximum output power in NTN.

EUTRA band	Class 3 (dBm)	Tolerance (dB)	Class 5 (dBm)	Tolerance (dB)
256	23	±2	20	±2
255	23	±2	20	±2

Source: Adapted from [5].

Table 8.2 NB-IoT/eMTC reference scenarios from TR 36.763.

Scenario	Orbit	Altitude (km)	Payload type	Coverage
A	GEO	35 786	Transparent	Earth-fixed beams
B	LEO	600, 1200		Earth-fixed beams
C	LEO	600, 1200		Earth-moving beams
D	MEO	10 000		Earth-moving beams

Source: Adapted from [5].

Table 8.3 Set-1 of satellite parameters for NB-IoT/eMTC.

Parameter	GEO, 35 786 km	LEO, 1200 km	LEO, 600 km
Satellite antenna pattern	Clause 6.4.1 in TR 38.811		
Central beam edge elevation	2.3°	26.3°	27.0°
Central beam center elevation	12.5°	30°	30°
Half Power Beam Width (HPBW)	0.4011°	4.4127°	
Equivalent satellite antenna aperture	22 m	2 m	
EIRP density	59 dBW/MHz	40 dBW/MHz	34 dBW/MHz
Maximum TX/RX gain	51 dBi	30 dBi	
G/T	16.7 dB/K	−12.8 dB/K	
Beam diameter at nadir	250 km	90 km	50 km

Source: Adapted from [5].

Table 8.4 Set-2 of satellite parameters for NB-IoT/eMTC.

Parameter	GEO, 35 786 km	LEO, 1200 km	LEO, 600 km
Satellite antenna pattern	Clause 6.4.1 in TR 38.811		
Central beam edge elevation	11.0°	22.2°	23.8°
Central beam center elevation	20°	30°	30°
Half Power Beam Width (HPBW)	0.7353°	8.8320°	
Equivalent satellite antenna aperture	12 m	1m	
EIRP density	53.5 dBW/MHz	34 dBW/MHz	28 dBW/MHz
Maximum TX/RX gain	45.5 dBi	24 dBi	
G/T	14 dB/K	−4.9 dB/K	
Beam diameter at nadir	450 km	190 km	90 km

Source: Adapted from [5, 7].

Table 8.5 Set-3 of satellite parameters for NB-IoT/eMTC [5].

Parameter	GEO, 35 786 km	LEO, 1200 km	LEO, 600 km
Central beam edge elevation	12.5°	30°	
Central beam center elevation	20.9°	46.05°	43.78°
Half Power Beam Width (HPBW)	0.7353°	22.1°	
Equivalent satellite antenna aperture	12 m	0.m	
EIRP density	59.8 dBW/MHz	33.7 dBW/MHz	28.3 dBW/MHz
Maximum TX/RX gain	45.7 dBi	16.2 dBi	
G/T	16.7 dB/K	−12.8 dB/K	
Beam diameter at nadir	459 km	470 km	234 km

Table 8.6 Set-4 of satellite parameters for NB-IoT/eMTC [5].

Parameter	LEO, 600 km
Central beam edge elevation	30°
Central beam center elevation	90°
Half Power Beam Width (HPBW)	104.7°
Equivalent satellite antenna aperture	0.097 m
EIRP density	21.45 dBW/MHz
Maximum TX/RX gain	11 dBi
G/T	−18.6 dB/K
Beam diameter at nadir	1700 km

Table 8.7 Set-5 of satellite parameters for NB-IoT/eMTC [5].

Parameter	MEO, 10 000 km
Central beam edge elevation	81.6°
Central beam center elevation	90°
Half Power Beam Width (HPBW)	6.5°
Equivalent satellite antenna aperture	1.5 m
EIRP density	45.4 dBW/MHz
Maximum TX/RX gain	28.1 dBi
G/T	3.8 dB/K
Beam diameter at nadir	1140 km

A summary of the link budget calibration results, based on the NTN channel model discussed in Chapter 3, is reported in [5].

For IoT NTN, Release-17 specified necessary enhancements to cope with the Doppler shift/variation and the delay variation which take maximum values in LEO-based IoT Deployment, and to cope with the large Round Trip Delay (RTD) delay, which takes maximum value in GEO deployment. Therefore, the IoT-NTN enhancements for LEO and GEO should be sufficient to support MEO scenario. And thereby, the parameter set for MEO provided in Table 8.7 is only for information/reference and evaluation/enhancements are mainly considered for GEO and LEO. These enhancements can be applicable for MEO.

8.2.3 NTN IoT Spectrum

E-UTRA operating bands for satellite access are given in the Table 8.8. Note that satellite bands are numbered in descending order from 256. The duplexing mode for satellite operation is FDD in all specified operating bands. As documented in [8], for operation in band 255 in USA and Canada when NS_02N is signaled, only channels positions which guarantee at least 90 kHz guard band from RF channel edge to the lower and upper limit of the band shall be used.

UE category M1 is designed to operate in the E-UTRA satellite access operating bands defined in Table 8.8 in both half duplex FDD mode and full-duplex FDD mode. Whereas, Category NB1 and NB2 UE operate in HD-FDD duplex mode.

Channel bandwidths for category M1, NB1, and NB2 in IoT NTN based access are given in Table 8.9.

Table 8.8 UTRA satellite access operating bands.

E-UTRA operating band	Uplink (UL) operating band BS receive UE transmit $F_{UL\_low} - F_{UL\_high}$	Downlink (DL) operating band BS transmit UE receive $F_{DL\_low} - F_{DL\_high}$	Duplex mode
256	1980 MHz–2010 MHz	2170 MHz–2200 MHz	FDD
255	1626.5 MHz–1660.5 MHz	1525 MHz–1559 MHz	FDD

Table 8.9 Channel bandwidth.

For category NB1 and NB2		For category M1	
Channel bandwidth $BW_{Channel}$ (kHz)	200	Channel bandwidth $BW_{Channel}$ (MHz)	1.4
Transmission bandwidth configuration N_{RB}	1	Transmission bandwidth configuration NRB	6
Transmission bandwidth configuration $N_{tone\,15kHz}$	12		
Transmission bandwidth configuration $N_{tone\,3.75kHz}$	48		

8.3 Enhancements for NB-IoT/eMTC Support in NTN

3GPP Relase-17 specified support of NB-IoT and eMTC over NTN. Several enhancements were introduced using NR NTN solutions already described as baseline. These include Enhancement on timing and synchronization, support of discontinuous coverage and mobility Management, and many other new features described in the following sections.

8.3.1 Timing and Frequency Pre-compensation

The network broadcasts information required for accessing an NTN cell in SIB31. These include ephemeris information and common Timing Advance (common TA) parameters. Before connecting to an NTN cell, IoT device shall acquire its GNSS position as well as the satellite ephemeris and common TA-related parameters.

To achieve uplink synchronization, the same principles from NR NTN (as described in Chapter 4) are reused for IoT NTN: Before performing RA and connecting to an NTN cell, the IoT device shall have valid GNSS position as well as the ephemeris and common TA. It shall autonomously pre-compensate the Timing Advance, and the frequency Doppler shift by considering the satellite position acquired through the satellite ephemeris, the common TA and its own position. In RRC connected mode, the UE shall continuously update the Timing Advance and frequency pre-compensation, but the UE is not expected to perform GNSS acquisition. The UE does not perform any transmissions due to outdated satellite ephemeris, common TA, or GNSS position based on timers. In connected mode, upon outdated satellite ephemeris and common Timing Advance, the UE re-acquires the broadcasted parameters and upon outdated GNSS position the UE moves to idle mode.

Timing Advance formula in Chapter 4 can be transposed to IoT-NTN with Ts used instead of Tc: Transmission of the uplink radio frame number i from the UE shall start $T_{TA} = \left(N_{TA} + N_{TA,offset} + N_{TA,adj}^{common} + N_{TA,adj}^{UE} \right) T_s$ seconds before the start of the corresponding downlink radio frame at the UE [9].

$$T_s = \frac{1}{(15\,000 \times 2048)} \text{ seconds}$$

$N_{TA,adj}^{common}$ is network-controlled common TA and may include any timing offset considered necessary by the network. It is derived from the higher-layer parameters *TACommon*, *TACommonDrift*, and *TACommon-DriftVariation* if configured, otherwise $N_{TA,adj}^{common} = 0$.

$N_{TA,adj}^{UE}$ is UE self-estimated TA to pre-compensate for the service link delay. It is computed by the UE based on UE position and serving satellite-ephemeris-related higher-layers parameters if configured, otherwise $N_{TA,adj}^{UE} = 0$.

The definition of the parameters related to timing and frequency pre-compensation are listed in Table 8.10. The TA components are illustrated in Figure 8.3 and the principles of frequency pre-compensation in IoT NTN are illustrated in Figure 8.4.

The UEs may be configured to report Timing Advance at initial access or in connected mode as specified in TS 36.321 [10]. In connected mode event-triggered reporting of the Timing Advance is supported.

For downlink synchronization in case of NB-IoT, the two LSB of the ARFCN is signaled in MIB for bands for which a 200 kHz channel raster is not supported, and the legacy 100 kHz raster is used. Otherwise, for bands for which a 200 kHz channel raster is supported, there is no signaling of ARFCN information in MIB.

Similar to NR NTN (refer to Chapter 4), the Doppler shift over the feeder link and any transponder frequency error for both Downlink and Uplink is compensated by the NTN GW and satellite-payload without any specification impacts in Release 17.

Uplink segmented transmission is supported for uplink transmission with repetitions. The UE shall apply UE pre-compensation per segment of UL transmission of PUSCH/PUCCH/PRACH for BL UEs and UEs in enhanced coverage and NPUSCH/NPRACH for NB-IoT from one segment to the next segment. Uplink segmented transmission NPRACH/NPUSCH for NB-IoT is not supported in GEO based on UE feature. The configuration

Table 8.10 SIB31(-NB) parameters for timing and frequency pre-compensation.

Parameter	Definition/description	Comment
orbitalParameters	Instantaneous values of the satellite orbital parameters. The signaled values are only valid for the duration as defined by ul-SyncValidityDuration and epochTime.	
stateVectors	Instantaneous values of the satellite state vectors. The signalled values are only valid for the duration as defined by ul-SyncValidityDuration and epochTime	
nta-Common	Network-controlled common TA	Unit of µs. Step of $32.552\,08 \times 10^{-3}$ µs. Actual value = field value * $32.552\,08 \times 10^{-3}$. If the field is absent, the UE uses the (default) value of 0.
nta-CommonDrift	Drift rate of the common TA	Unit of µs/s. Step of 0.2×10^{-3} µs/s. Actual value = field value * 0.2×10^{-3}. If the field is absent, the UE uses the (default) value of 0.
nta-CommonDrift Variation	Drift rate variation of the common TA	Unit of µs/s². Step of 0.2×10^{-4} µs/s². Actual value = field value * 0.2×10^{-4}. If the field is absent, the UE uses the (default) value of 0.

(Continued)

Table 8.10 (Continued)

Parameter	Definition/description	Comment
epochTime	Epoch time of the satellite ephemeris data and common TA parameters. epochTime is the starting time of a DL subframe indicated by startSFN and startSubframe. For serving cell, the startSFN indicates the current SFN or the next upcoming SFN after the frame where the message indicating the epochTime is received. If the field is absent, the UE uses the starting time of the DL subframe corresponding to the end of the SI window during which the SI message carrying SIB31(-NB) is transmitted.	The reference point (RP) for epoch time of the serving satellite ephemeris and Common TA parameters is the uplink time synchronization RP.
ul-SyncValidity Duration	Validity duration of the satellite ephemeris data and common TA parameters, i.e., maximum time duration (from epochTime) during which the UE can apply the satellite ephemeris without acquiring new satellite ephemeris	Unit in second. Value *s5* corresponds to 5 seconds, value *s10* corresponds to 10 seconds, and so on.

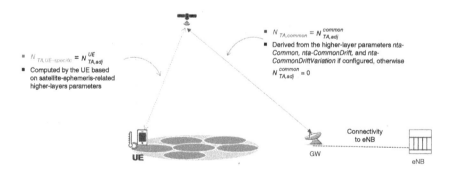

$$T_{TA} = (N_{TA} + N_{TA,UE-specific} + N_{TA,common} + N_{TA,offset}) \times T_s$$

Figure 8.3 UE-specific TA and Common TA.

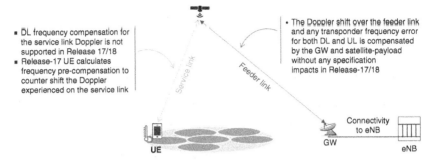

- DL frequency compensation for the service link Doppler is not supported in Release 17/18
- Release-17 UE calculates frequency pre-compensation to counter shift the Doppler experienced on the service link

- The Doppler shift over the feeder link and any transponder frequency error for both DL and UL is compensated by the GW and satellite-payload without any specification impacts in Release-17/18

Connectivity to eNB

GW

UE

eNB

Figure 8.4 Frequency pre-compensation in IoT-NTN.

of uplink transmission segment is indicated on SIB for initial access and can be re-configured by RRC signaling.

System Information Block Type31 (SIB31) is a new SIB introduced for IoT NTN support. It contains satellite assistance information for the serving cell. It is only signaled in an NTN cell. Parameters related to timing and frequency pre-compensation broadcast in SIB31(-NB) are listed in Table 8.10.

8.3.1.1 Uplink Synchronization Validity Duration

A validity duration for UL synchronization (for both satellite ephemeris data and common TA parameters) configured by the network is specified: The parameter **ul-SyncValidityDuration** broadcast in SIB31(-NB) (see Table 8.10) is used to indicate this validity duration, i.e., maximum time duration (from epochTime, see definition in Table 8.10) during which the UE can apply the satellite ephemeris without acquiring new satellite ephemeris.

8.3.1.2 GNSS Operation in IoT NTN

UE power efficiency is an important factor that should be taken into account for IoT NTN design. As the uplink time and frequency synchronization mechanisms is dependent on GNSS usage, GNSS position fix impact on UE power consumption in IoT NTN is a critical issue that needed to be tackled. Battery life analysis with GNSS position fix every UL transmission was conducted during Relase-17 study on NB-IoT/eMTC support for NTN. Both GNSS power consumption (in case of integrated GNSS and IoT module and separate GNSS module and IoT Module) and GNSS position Time To First Fix (TTFF) impact on UE power consumption were investigated considering the battery life methodology reported in TR 45.820 [7, 11]. Extensive numerical results for the battery life are reported in Annex C of [5]. It was observed that the GNSS acquisition time is 30 seconds with cold start (the terminal shall search for any satellite without prior

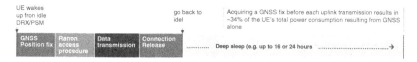

Figure 8.5 Short, sporadic transmissions for IoT over NTN.

reliable information), a few seconds with warm start (the terminal has rough estimates of the position, timing, and almanac data so as to acquire the signals from the required satellites), and up to 2 seconds in hot start (the terminal has accurate position, timing, and almanac data to start from).

As reported in [5], under the studied scenario for short and sporadic connections as shown in Figure 8.5, it was observed that, acquiring a GNSS fix before each uplink transmission results in ∼34% of the UE's total power consumption resulting from GNSS alone. Moving IoT devices such as tracking devices using short and sporadic connection may suffer from GNSS position fix impacting the UE power consumption. However, fixed IoT devices such as a smart meter or a water meter, may be able to save power by having a much more relaxed (e.g. once a week, or once a month, depending on the setting) GNSS position fix. For a long connection employing connected mode DRX (with a Discontinuous Reception [DRX]cycle of ∼10 seconds) as illustrated in Figure 8.6, under the studied scenario reported in [5], it was observed that a GNSS fix before every uplink transmission consumes approximately 45% of the UE's total available energy without additional enhancements.

With the above in mind, 3GPP Release-17 specified for sporadic short transmission in RRC CONNECTED mode a validity of a GNSS position fix and details of acquiring a GNSS position fix: For sporadic short transmission, the idle UE wakes up from idle DRX/Power Saving Mode (PSM), access the network, perform uplink and/or downlink communications for a short duration of time, and go back to idle. Before accessing the network, the UE acquires GNSS position fix and does not need to re-acquire a GNSS position fix for the transmission of the packets. If GNSS becomes outdated, UE in RRC_CONNECTED should go back to idle mode and re-acquire a GNSS position fix.

The UE autonomously determines its GNSS validity duration (i.e. **gnss-validityDuration**) and reports information associated with this valid duration to the network via RRC signaling. The **gnss-validityDuration**

Figure 8.6 Long connection with connected mode DRX for IoT over NTN.

indicates the remaining GNSS validity duration in the UE: Value s10 corresponds to 10 seconds, s20 corresponds to 20 seconds, and so on. Value min5 corresponds to 5 minutes, value min10 corresponds to 10 minutes, and so on. GNSS-ValidityDuration is reported by the UE in RRCConnectionSetup Complete-NB message, which is used to confirm the successful completion of an RRC connection establishment. And in RRCConnectionReconfigurationComplete, RRCConnectionReestablishmentComplete-NB, RRC ConnectionResumeComplete-NB messages, which are used respectively to confirm the successful completion of an RRC connection reconfiguration, to confirm the successful completion of an RRC connection re-establishment, and to confirm the successful completion of an RRC connection resumption.

The duration of the short transmission is not larger than the validity timer for UL synchronization, i.e., **ul-SyncValidityDuration**. With a GNSS position fix that can be assumed to be valid for some period of time, i.e., **gnss-validityDuration**, the following apply for UE in RRC_CONNECTED: TA error due to UE velocity satisfies the timing synchronization requirements and Doppler shift error due to UE velocity satisfies the frequency synchronization requirement. Thereby, satellite ephemeris and Common TA parameters if indicated and read on SIB are valid for the duration of sporadic short transmission in RRC_CONNECTED.

3GPP Release-18 introduced further enhancements to optimize the GNSS operation with sparse use of GNSS and power efficiency for long-term connection (compared to Release-17). At the time of writing this book, the 3GPP normative work is not yet completed. However, the following summaries in bullets the main enhancements under the sub agenda item on improved GNSS operations for IoT NTN:

- IoT NTN UE may need to re-acquire a valid GNSS position fix in long connection time
- UE reports additional GNSS assistance information, including GNSS position fix time duration for measurement and GNSS validity duration:
  - UE reports GNSS position fix time duration for measurement at least during the initial access stage
  - UE reports only one GNSS position fix time duration for GNSS measurement at least when moving to RRC connected state
  - In connected mode, UE may report GNSS validation duration with MAC CE
  - For the GNSS measurement gap aperiodically triggered with MAC CE, the duration for the GNSS measurement gap can be configured by eNB.
    - The gap duration is equal to the latest reported GNSS position fix time duration for measurement when the duration for GNSS measurement gap is not included in the configuration by eNB.

- UE reports one GNSS position fix time duration for GNSS measurement via a 4-bit field with component values [1,2,3,4,5,6,7,13,19,25,31]
  - The UE is not required to transmit or receive any channel/signal within the aperiodic GNSS measurement gap duration before the UE reacquires GNSS successfully
  - Release-18: Support eNB to trigger UE to make GNSS measurement in aperiodic manner:
    - For GNSS measurement in RRC connected, if eNB triggers in aperiodic way connected UE to make GNSS measurement, UE can re-acquire GNSS position fix with a gap
    - The UE may re-acquire GNSS autonomously (when configured by the network) if UE does not receive eNB trigger to make GNSS measurement

8.3.2 Timing Relationship Enhancements

Downlink and uplink timings are frame-aligned at the uplink time synchronization Reference Point (RP). To accommodate the long propagation delays in NTN, the timing relationships are enhanced by the support of two scheduling offsets: Koffset and Kmac.

The scheduling offset Koffset is used to allow the UE sufficient processing time between a downlink reception and an uplink transmission as specified in TS 36.213 [12]. The following timing relationship in IoT NTN enhanced with the scheduling Koffset are:

- For NB-IoT, on receiving UL grant on DCI format N0 in subframe n, NPUSCH Format 1 is transmitted with a delay of Koffset.
- For NB-IoT, on receiving an NPDSCH with a RAR message that ends in subframe n, the corresponding Msg3 is transmitted on NPUSCH format 1, with a delay of Koffset.
- For NB-IoT, a UE upon detection of an NPDSCH transmission for which it should provide a ACK/NACK feedback, shall transmit the HARQ ACK/NACK with a delay of Koffset
- For NB-IoT, on receiving a timing advance command ending in DL subframe n, the corresponding adjustment of the uplink transmission timing by the received time advance shall be delayed by Koffset.
- In IoT NTN, for a RA procedure initiated by an N/MPDCCH order, the UE shall delay the transmission of the RA preamble by Koffset.
- For emTC, on receiving an UL grant via MPDCCH that ends in DL subframe n, PUSCH is transmitted with a delay of Koffset.
- For emTC, on receiving a RAR in a PDSCH that ends in subframe n, PUSCH for Msg3 is transmitted with a delay of Koffset.

- For eMTC, when an MPDCCH ending in subframe n activates UL SPS, the time of the first subframe in which the UE is allowed to transmit SPS-PUSCH is delayed by Koffset.
- For eMTC, on reception of a PDSCH ending in subframe n, the corresponding HARQ-ACK feedback on PUCCH is transmitted with a delay of Koffset.
- For eMTC, the ending time for DL physical resources forming a CSI reference resource set is advanced by Koffset.
- For eMTC, for an MPDCCH received in subframe n that triggers aperiodic SRS transmission, SRS is transmitted with a delay of Koffset.
- For eMTC, on receiving a timing advance command ending in subframe n, the corresponding adjustment of the uplink transmission timing by the received time advance shall be delayed by Koffset.
- For eMTC in IoT NTN, if the UE determines that a preamble retransmission is necessary, the choice of a suitable preamble retransmission subframe shall be delayed by Koffset.

For IoT NTN, with respect to the configuration, indication, and update of K_Offset, the mechanisms concluded in NR-NTN and described in Chapter 4 are taken as baseline: For IoT NTN, Release-17 specified cell-specific Koffset (**CellSpecificKoffset**) configuration for use during initial access. **CellSpecificKoffset** is signaled in system information. Further, the network may provide and update a UE-specific K_offset using MAC CE. Which means that the use of UE-specific Koffset in CONNECTED mode is also supported.

For a BL/CE UE, if the UE is configured with the higher-layer parameter **CellSpecificKoffset**,

$$K_{offset} = K_{cell\_offset} - K_{UE\_offset}$$

where

K_{cell_offset} is the parameter **CellSpecificKoffset** provided by higher layers, and

K_{UE_offset} is the parameter UE Specific Koffset provided by higher layers, otherwise $K_{UE\_offset} = 0$

otherwise,

$K_{offset} = 0, K_{cell\_offset} = 0.$

In IoT NTN, the Koffset value signaled in system information (i.e. **Cell-SpecificKoffset**) is always used for NPDCCH and MPDCCH ordered NPRACH and PRACH timing relationships, respectively.

For IoT NTN, the unit of K_offset is subframe based on a 15 kHz subcarrier spacing (i.e. 1 ms). The unit of Koffset when subcarrier spacing is 3.75 kHz is 1 ms.

As outlined in Chapter 4, Kmac may be provided by the network when downlink and uplink frame timing are not aligned at eNB. This offset is used to delay the application of a downlink configuration indicated by a MAC CE received on NPDSCH/PDSCH. Further, the following timing relationships in IoT NTN are enhanced with Kmac:

- For NB-IoT, if the UE has initiated an NPUSCH transmission using pre-configured uplink resources ending in subframe n, the UE shall start or restart to monitor the NPDCCH from DL subframe $n + 4 + K\_mac$ (where K_mac is defined as in NR-NTN).
- For emTC, if the UE has initiated a PUSCH transmission using pre-configured uplink resources ending in subframe n, the UE shall start or restart to monitor the MPDCCH from DL subframe $n + 4 + K\_mac$ (where K_mac is defined as in NR-NTN).

For IoT NTN, the information of K_mac is carried in system information. The unit of K_mac is subframe based on a 15 kHz subcarrier spacing (i.e., 1 ms). And when subcarrier spacing is 3.75 kHz, the unit of K_mac is 1 ms.

Parameters related to timing relationship enhancements broadcast in SIB31(-NB) are listed in Table 8.11.

Table 8.11 SIB31(-NB) parameters for timing relationship enhancements.

Parameter	Definition/description	Comment
k-Offset	Scheduling offset used in the timing relationships in NTN	Unit in ms
k-Mac	Scheduling offset used when downlink and uplink frame timing are not aligned at the eNB	Unit in ms. If the field is absent, the UE uses the (default) value of 0.

8.3.3 Discontinuous Coverage and Assistance Information

As a satellite moves on a specified orbit, in case of an NGSO satellite, the satellite beam(s) coverage area may move and cover different portions of a geographical area due to the orbital movement of the satellite as illustrated in Figure 8.7. Thereby, an IoT device located in the concerned geographical area may experience a situation of discontinuous coverage, due to, e.g., a sparse satellite constellation deployment.

With the above in mind, it would be beneficial to enable UEs in RRC_IDLE to predict upcoming satellite fly-over periods and save power during periods of no coverage. The network may broadcast assistance information relating to the serving satellite and other satellites of the constellation. The broadcast assistance information includes SGP4 ephemeris elements based on the Two-Line Elements (TLE) sets industry standard. Additional assistance information, such as the start time of upcoming satellite coverage, footprint parameters, and cell radius, may also be optionally broadcast by the network.

It is optional for an UE camped on NTN cell to support discontinuous coverage as specified in TS 36.304 [13]. Further, predicting out-of-coverage and in-coverage periods is up to UE implementation. When out of coverage, the UE is not required to perform Access Stratum (AS) functions.

The SIB Type 32 contains satellite assistance information for prediction of discontinuous coverage. This is a new SIB introduced in Release-17 and is only signaled in an NTN cell. The content of SIB32 is shown in Table 8.12.

In the Core Network, discontinuous coverage is handled by means of Tracking Area- or RAT-specific configuration of the Mobility Management Entity (MME) such that the MME is able, via existing functionality (namely periodic TAU timer, mobile reachable timer, implicit detach timer and high latency communication), to ensure that when the UE is unreachable,

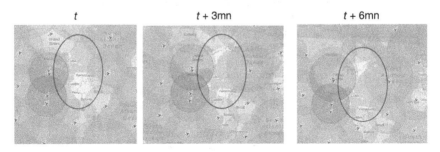

Figure 8.7 Discontinuous coverage in NTN IoT.

Table 8.12 SIB32 content.

Parameter	Definition/description	Comment
footprintInfo	Satellite footprint. E-UTRAN may configure *elevationAngles* and/or *radius* for earth-moving cells. E-UTRAN may configure *referencePoint* and *radius* for quasi-earth fixed cells.	
elevationAngleLeft, elevationAngleRight	Leftmost and rightmost (with reference to the satellite direction) elevation angle.	Unit in degree. Step of 5 degree. Actual value = field value * 5. If the field elevationAngleLeft is absent, the leftmost elevation angle is equal to the value of field elevationAngleRight
latitude	Latitude of the reference point (RP). Unit in degree.	Step of 360/262 144 degree. Actual value = field value * (360/262 144).
longitude	Longitude of the RP. Unit in degree.	Step of 360/262 144 degree. Actual value = field value * (360/262 144).
radius	Distance between the RP and the edge of the satellite or beam coverage.	Unit in km. Step of 10 km. Actual value = field value * 10.
serviceInfo	Information on when the satellite will provide coverage. E-UTRAN always configures *tle-EphemerisParameters* for a satellite with earth moving cell(s) and always configures *t-ServiceStart* for a quasi-earth fixed cell.	
tle-Ephemeris Parameters	Mean values of the satellite orbital parameters based on the TLE set format for estimating in-coverage and out-of-coverage periods for a satellite with earth-moving cell(s).	
t-ServiceStart	Time information on when the incoming satellite is going to start serving the area for quasi-earth fixed cells.	

(i) the UE does not trigger NAS transaction or detach from the network and (ii) mobile-terminated data destined to the UE can be stored in the network.

8.3.4 Mobility Management

For IoT NTN with moving cells, in order to ensure that each TA is Earth-stationary, even if the radio cells are moving across the Earth's surface, the E-UTRAN may change the TAC values that are broadcast in a cell's system information as the cell moves. An example of Tracking Area handling in CIoT over satellite access is shown in Figure 8.8.

For the Mobility Management in ECM-IDLE, the network may broadcast more than one Tracking Area Code (TAC) per PLMN in a cell in order to reduce the signaling load at cell edge in NTN, in particular, for Earth-moving cell coverage. The AS layer indicates all received TACs for the selected PLMN to the NAS layer. The network may update the UEs upon TAC removal. By UE implementation, UEs may also check whether a TAC has been removed from the TACs broadcast by the network.

At the NAS layer, the UE need not trigger a Tracking Area Update due to mobility reasons, if any of the broadcast TAC(s) in the cell where the UE is located is part of the UE's Tracking Area List.

For quasi-Earth-fixed cells, timing information on when the cell is going to stop serving the area may be broadcast by the network. This may be used by the UE to start measurements on neighbor cells before the broadcast stop time of the serving cell, while the exact start of the measurements is up to UE implementation.

For the Mobility Management in ECM-CONNECTED, Radio link failure and RRC connection re-establishment are supported in NTN. To enable

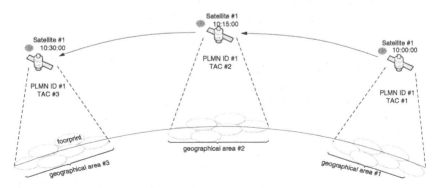

Figure 8.8 Tracking Area handling in Cellular IoT over satellite access.

mobility in NTN, the network provides target cell satellite parameters needed to access the NTN cell in the handover command. Conditional handover is supported for BL UEs and UEs in enhanced coverage.

Different RAT types are introduced that allow distinction by the Core Network between existing terrestrial accesses and new non-terrestrial accesses as well as, among non-terrestrial accesses, between the different types of satellite constellations (LEO, MEO, GEO, and OTHERSAT) and radio access type (i.e. WB-EUTRAN, NB-IoT, and LTE-M). This allows the Core Network nodes and the Home Subscriber Server (HSS) to identify the access a UE is using such that they are able to adjust their behavior and that of the UE accordingly (e.g., setting of NAS timers, and determination and enforcement of access restrictions).

8.3.5 Feeder-link Switchover

As defined in [14], a feeder link switchover is the procedure where the feeder link is changed from a source NTN GW to a target NTN GW for a specific NTN payload. The feeder link switchover is a Transport Network Layer procedure. The NTN Control function determines the point in time when a feeder link switchover between two eNBs is performed. For BL UEs and UEs in enhanced coverage, the transfer of the affected UE(s)' context between the two eNBs at feeder link switchover is performed by means of either S1-based or X2-based handover, and it depends on the eNBs' implementation and configuration information provided to the eNBs by the NTN Control function. More details on Feeder-link switchover procedure could be found in Section 6.9.

8.3.6 Network-interfaces Signaling Aspects

The Cell Identity in NTN corresponds to a fixed geographical area identified by a Mapped Cell ID, irrespective of the orbit of the NTN payload or the type of the service link. For a BL UE or a UE in enhanced coverage, the Cell Identity included within the target identification of the handover messages allows identifying the correct target cell.

The mapping between Mapped Cell IDs and geographical areas is configured in the RAN and the Core Network (e.g., pre-configured depending on operator's policy, or based on implementation). For a BL UE or a UE in enhanced coverage or a NB-IoT UE that supports S1-U data transfer or UP CIoT EPS optimization, the eNB is responsible for constructing the Mapped Cell ID based on the UE location information received from the UE, if available. The User Location Information (ULI) may enable the MME to determine whether the UE is allowed to operate at its present location.

Pre-configuration of special mapped cell identifiers may be used to indicate areas outside the serving PLMN's country.

The eNB reports the broadcasted TAC(s) of the selected PLMN to the MME. In case the eNB knows the UE's location information, the eNB may determine the TAI the UE is currently located in and provide that TAI to the MME.

8.3.7 MME(Re-)Selection by eNB

For an RRC_CONNECTED UE, when the eNB is configured to ensure that the BL UE or the UE in enhanced coverage is using an MME that serves the country in which the UE is located. If the eNB detects that a BL UE or a UE in enhanced coverage is in a different country from that served by the serving MME, it should perform an S1 handover to change to an appropriate MME or initiate a UE Context Release Request procedure toward the serving MME (in which case the MME may decide to detach the UE).

For an RRC_CONNECTED NB-IoT UE, when the eNB is configured to ensure that the NB-IoT UE is using an MME that serves the country in which the UE is located. If the eNB detects that the UE is in a different country than that served by the serving MME, it should initiate a UE Context Release Request procedure toward the serving MME (in which case the MME may decide to detach the UE).

8.3.8 Verification of UE Location

As specified in TS 23.401 [15], the network may, according to regulatory requirements, need to enforce that the PLMN selected by the UE is allowed to operate in the geographical location where the UE is located. To this end, the MME may invoke the ULI procedure during Mobility Management and Session Management procedures in order to determine the UE location. If the MME is able to determine with sufficient accuracy that it is not allowed to operate in the UE location it may reject and/or detach the UE.

8.3.9 O&M Requirements

The NTN-related parameters shall be provided by O&M to the eNB providing non-terrestrial access, as specified in TS 38.300 for NR NTN and described in section 6.10 (Chapter 6).

8.3.10 Other NAS Protocol Aspects

Enhancements to NAS signaling allow the UE to register to EPS core network using satellite E-UTRAN radio access technology. UICC-ME interface is extended to support network selection over satellite access and allowing to prioritize networks offering satellite access. EPS NAS re-transmission timers are extended to support longer propagation delays and response times due to extended distance between peer entities when satellite access is used. The UE supporting satellite E-UTRAN access supports also GNSS and potential uplink signaling delays to be considered in UE and network NAS implementations.

References

1 C. Amatetti, "NB-IoT via non terrestrial networks", Ph. D. Dissertation, Alma Mater Studiorum Università di Bologna, June 2023. doi: 10.48676/unibo/amsdottorato/11058.

2 RP-210868: "New Study WID on NB-IoT/eTMC support for NTN", Mediatek, RAN#91-e, March 2021.

3 3GPP TR 38.821 V16.1.0: "Solutions for NR to support non-terrestrial networks (NTN) (Release 16)", May 2021.

4 RP-210908: "Solutions for NR to support non-terrestrial networks (NTN)", Rapporteur (Thales), RAN#91-e, March 2021.

5 3GPP TR 36.763 V17.0.0: "Study on Narrow-Band Internet of Things (NB-IoT)/enhanced Machine Type Communication (eMTC) support for Non-Terrestrial Networks (NTN) (Release 17)", June 2021.

6 3GPP TR 38.811 V15.4.0: "Study on New Radio (NR) to support non-terrestrial networks (Release 15)", September 2020.

7 3GPP TR 45.820 V13.1.0: "Cellular system support for ultra-low complexity and low throughput Internet of Things (CIoT)", December 2015.

8 TS 36.102: "Evolved Universal Terrestrial Radio Access (E-UTRA); User Equipment (UE) radio transmission and reception for satellite access". September 2023.

9 TS 36.211, E-UTR: "Physical channels and modulation (Release 17)".

10 TS 36.321: "Evolved Universal Terrestrial Radio Access (E-UTRA); Medium Access Control (MAC) protocol specification". September 2023.

11 A. Guidotti et al., "Architectures, standardisation, and procedures for 5G Satellite Communications: a survey", *Computer Networks*, vol. 183, p. 107588, December 2020.

12 3GPP TS 36.213: "Evolved Universal Terrestrial Radio Access (E-UTRA); Physical layer procedures". September 2023.

13 TS 36.306: "Evolved Universal Terrestrial Radio Access (E-UTRA); User Equipment (UE) radio access capabilities". March 2023.

14 TS 36.300: "Evolved Universal Terrestrial Radio Access (E-UTRA) and Evolved Universal Terrestrial Radio Access Network (E-UTRAN); Overall description; Stage 2". July 2023.

15 TS 23.401: "General Packet Radio Service (GPRS) enhancements for Evolved Universal Terrestrial Radio Access Network (E-UTRAN) access". September 2023.

9

Release 18 and Beyond

9.1 NTN in the Evolving Context of 5G, Beyond 5G and 6G

Prior to the advent of 5G, satellite and mobile networks were designed independently from one another and were addressing separate user markets. With the 5G standard, Non-Terrestrial Networks (NTN) have been introduced as added network components to cellular networks with the primary objective to extend the service coverage in unserved or underserved areas. As depicted in Figure 9.1, the definition of the 6G system offers an opportunity to unify natively both "terrestrial" network (TN) and NTN components at system architecture and protocol level allowing a smart combination of both. In this chapter, the added value and characteristics of the NTN component of 5G system are recalled and its evolution in the context of beyond 5G systems including 6G is highlighted.

9.2 Non-Terrestrial Networks and 5G

9.2.1 3GPP Standardization Status

As part of 3GPP Release 17, a set of enhancing features enabling 4G (Narrowband-Internet of Things, NB-IoT, and enhanced Machine Type Communications, eMTC) and 5G New Radio (NR) radio protocols to be used in satellite networks have been defined. Moreover, the Radio Frequency (RF) and Radio Resource Management (RRM) performance applicable to satellite access nodes and NTN capable user equipment (UE) have been defined for Mobile Satellite Service (MSS) allocated bands commonly named L and S bands.

5G Non-Terrestrial Networks: Technologies, Standards, and System Design, First Edition.
Alessandro Vanelli-Coralli, Nicolas Chuberre, Gino Masini, Alessandro Guidotti, and Mohamed El Jaafari.
© 2024 The Institute of Electrical and Electronics Engineers, Inc. Published 2024 by John Wiley & Sons, Inc.

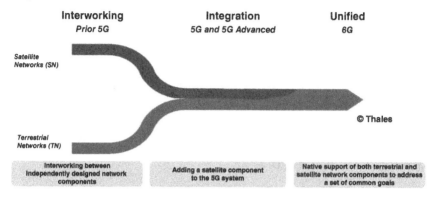

Figure 9.1 Interactions between "terrestrial" and non-terrestrial networks before and beyond 5G.

These NTN features address specific technical issues of satellite networks compared to cellular networks, namely:

- Propagation channel and link budget constraints;
- Extended/variable propagation delays and Doppler;
- Wider and possibly moving radio cells;
- Mobility between TN and NTN;
- Reliable determination of UE location to support regulatory services, such as lawful intercept, emergency calls, or public warning services.

The NTN enhancements defined in Releases 17/18 by the different 3GPP Radio Access Network (RAN) working groups are listed in Figure 9.2.

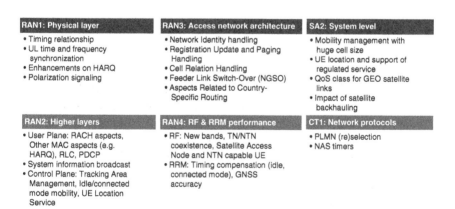

Figure 9.2 Rel-17 NTN impacts on 5G NR/NG-RAN specifications.

These enhancements were crafted so as to minimize impacts on both terminals and RAN while supporting the widest range of satellite network deployment scenarios, i.e.:

- Any Non Geosynchronous Orbit (NGSO) satellite at 600 km altitude and above as well as any Geosynchronous Orbit (GSO) satellite, High Altitude Platform Stations (HAPs);
- Any beam size from few to thousands of kilometers diameter;
- Earth-fixed and Earth-moving beams;
- Any band below 7 GHz and above 10 GHz;
- Any terminal: Commercial smartphone and Internet of Things (IoT) devices with omnidirectional antenna and 200 mW transmit power up to higher performance terminals (antenna gain and transmit power).

This is in line with the principle set forth by Richard Locke in [1]: "Future satellite air interfaces must be universal to support the economy of scale." Release 18 is currently being defined with the introduction of:

- Enhancements to the NR radio protocol, in order, to support satellite networks operating above 10 GHz and addressing mobile and nomadic Very Small Aperture Terminals (VSAT); to allow the verification of the Global Navigation Satellite System (GNSS) coordinates determined by the UE and to optimize mobility procedures in both idle and connected modes;
- Enhancements to NB-IoT and eMTC radio protocols, to optimize mobility procedures and to improve the support of small constellations providing discontinuous service over a given area.

The reference scenarios reported in Figure 9.3 are being considered for 5G NTN, whose differences originate from the terminals they target.

The standardization effort is then meant to enable the integration of satellite communications with the 3GPP ecosystem, and to support:

- Economy of scale associated with a global market perspective and a wide set of vendors at chipset, terminal, network, and service level;

	Release 17		Release 18
	Direct connectivity (<7 GHz)		Indirect connectivity(above 10 GHz)
Targeted terminals	IoT devices	handset (smart-phones) and car/drone mounted devices	VSAT and/or ESIM
Service	Narrowband	Wideband	Broadband
	hundreds of kbps	few Mbps	hundred Mbps
3GPP Radio interfaces	4G NB-IoT/eMTC	5G New Radio	5G New Radio
Example of applications	*Professional: utilities (smart grids, water distribution, oil & gas), agriculture*	*Consumer market Professional markets: Automotive, public safety, utilities, agriculture, Defense*	*Professional markets: Telco (e.g. Backhaul), IPTV service providers, Satellite News Gathering, Transport (aeronautical, maritime, railway), public safety, defense*

Figure 9.3 Reference scenarios for 5G NTN.

- Multi-vendor interoperability which all users (e.g., Public safety, transport, automotive, drone, and defense) of satellite communication systems demand to avoid locking with a proprietary satellite communication systems;
- Continuous service and capability innovation beyond 5G and 6G;
- Backward compatibility between the generations.

3GPP NTN-compliant satellite network infrastructures will be able to support:

- Native 5G features such as slicing, Quality of Service (QoS), security, energy saving, and private networks;
- Global service continuity through mobility between NTN and TN;
- Improved Quality of Experience (QoE) and reliability through multi-connectivity between terrestrial and satellite access GSO/NGSO;
- Spectrum coexistence (adjacent band) of satellite and mobile systems.

9.2.2 Industrial Projects Based on Rel-17 and Rel-18

9.2.2.1 Direct Connectivity to Smartphones

Several projects considering the provision of direct satellite connectivity to smartphones have been disclosed in 2022, all of which focus on the development of fast-track solutions to offer a service as soon as possible and benefit from a first-mover advantage. These projects:

- Offer limited service (e.g., short messaging service capability);
- Reuse key building blocks such as legacy terminals, legacy satellite services, or in-orbit space segment resources or services based on in-orbit NGSO or GSO constellation.

Service to legacy smartphones enforces

- The reuse of legacy Long Term Evolution (LTE) radio interfaces, NR, or even Global System for Mobile communications (GSM), with a proprietary implementation on the network side to mitigate the specifics of the space segment (e.g., Doppler, latency, and radio cell pattern);
- Adaptations of regulatory framework and advanced techniques to address inter-system interference issues stemming from the sharing of spectrum with existing mobile systems.

These incremental approaches could achieve a fast service roll-out, yet they face hurdles in terms of integration with existing mobile systems. Migration issues to upgrade the service capabilities are also to be expected.

In contrast, a native 3GPP NTN-compliant satellite network deployed in MSS bands requiring the upgrade of smartphones, a new ground network infrastructure (as well as possibly a new space segment) may take a bit longer to develop and deploy but it will support the necessary scalability for wideband services (a few Mbps) that users are expecting.

9.2.2.2 Direct Connectivity to IoT Devices

The limited revenue perspective associated with IoT applications puts a challenge on the capital expenditure on the network infrastructure, including space and ground segment. This challenge is addressed by:

- Leveraging existing geostationary satellites operating in MSS bands. In this case, the overall development/deployment schedule is then mostly constrained by the development and certification of the IoT devices themselves;
- Small, innovative non-geostationary satellite constellation operating in MSS bands and providing a lacunary coverage. This is made possible with the implementation of store and forward scheme compensating for the discontinuous coverage. The development/deployment schedule is then mostly driven by the development and deployment of the satellite(s) and the ground network.

Both cases seem compatible with a commercial service opening in 2024–2025.

9.2.2.3 Connectivity to Cell

Legacy broadband satellite networks operating above 10 GHz provide connectivity to access points (e.g., Wi-Fi) or backhaul service to base stations in remote locations or on-board moving platforms such as vessels, airplanes, or trains. Given that the 3GPP NTN standard was unavailable until 2022, it has not been considered for the networks currently being developed. Thus:

- Networks based on GSO satellites are typically using radio protocols leveraging DVB-S2X/RCS2 specifications which do not ensure multi-vendor interoperability.
- Networks based on NGSO mega-constellations are typically using proprietary radio protocols.

Replacing the above networks with NTN 3GPP-compliant satellite networks will address three pains reported by telecom operators:

- QoS is easier to manage end to end, including across the satellite network link;

- Breaches introduced by satellite network links in the mobile network security architecture can be prevented;
- RRM can be coordinated between local access points and the satellite network links.

9.3 Toward 6G and Non-Terrestrial Networks

9.3.1 6G System Versus 5G System

The advent of the 5G system was triggered by the combination of exponential growth in mobile network traffic, driven by media and entertainment, and an intent to address vertical users and their specific applications. Vertical stakeholders require new or improved service capabilities compared to consumers. This includes low latency, high reliability, density of connected devices, or positioning.

Although 6G network requirements remain in their very infancy, ITU-R WP 5D recently initiated a document putting together the International Mobile Telecommunications (IMTs) vision for 2030 and beyond [2]. It relies on the outcome of the ITU-T Focus Group Technologies for Network 2030 (FG-NET-2030), which defined a set of preliminary target services for 6G. The main vertical drivers highlighted include Virtual Reality (VR), Augmented Reality (AR), digital twinning, immersive communications and multi-sensory interactions (e.g., tactile/haptic Internet), Integration of sensing and communication, collaborative robots, autonomous driving of vehicles and drones, cognition, and connected intelligence [3, 4].

6G is expected to create a fully connected world, where the physical world is digitalized with high detail so as to be analyzed and acted upon. In this vision, the network would thus support interactions between the human, digital, and physical domains by devices embedded everywhere, by providing the infrastructure and the intelligence of the digital domain. Thus, 6G would enhance human and machine-type communications with increased performance, enable new services based on advanced positioning and/or sensing, and support the evolution of network in its ability to provide enhanced trust and coverage, while addressing spectrum and energy scarcity. As a result, humans would evolve in a cyber-physical continuum connecting bodies and intelligence. Achieving this vision would imply that 6G brings a lot more than just extremely fast mobile connectivity.

When discussing 6G, one should distinguish the capabilities of 5G, which will be enhanced from the new capabilities that will be enabled. For both cases, the NTN component has a specific role.

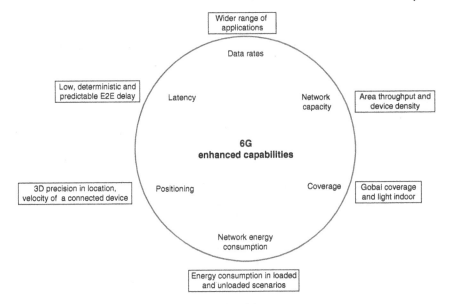

Figure 9.4 Enhanced capabilities compared to 5G.

The NTN component will contribute to enhance 5G capabilities as follows (see Figure 9.4):

- Data rates and network capacity in areas not covered by TN: an increase by at least one or two orders of magnitude to sustain the exponential increase of the traffic is expected;
- Coverage: Direct connectivity from NTN will be made possible in light indoor conditions;
- Energy consumption of the network infrastructure: Significant reduction is expected, thanks to smart routing between NTN and TN nodes;
- Positioning: Higher accuracy and lower acquisition time in areas beyond TN coverage will be achieved;
- Latency: Deterministic and lower round-trip delays will be enabled in areas beyond TN coverage, for example, with the deployment of interconnected network nodes on very low orbiting satellites, HAPs, and drones.

The NTN component will be expected to contribute to the new capabilities of 6G as follows (see Figure 9.5):

- Service availability: True seamless service continuity will be enabled to mask any hole (no perceived service interruption, no data loss) of the TN component coverage;

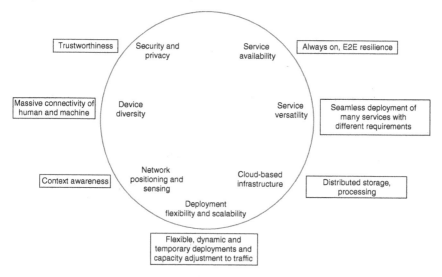

Figure 9.5   New capabilities enabled by 6G.

- Service versatility: Thanks to a smart combination of TN and NTN components, it will be possible to adjust the manifold characteristics of the service (e.g., in terms of latency reliability, bandwidth, and connectivity density) to the targeted needs;

- Cloud-based infrastructure: Each NTN flying platform (Satellite, HAPS, or even drones) will be able to embark network nodes with edge computing and storage capabilities that can be exploited to provide faster response time for certain applications;

- Deployment flexibility and scalability in areas not covered by TNs: while satellite constellation typically provide regional or global coverage, HAPS and drones may be added to reinforce the capacity in some specific areas and hence adjust the network capacity to the geographical traffic distribution;

- Network positioning and sensing: The network will be able to support trusted positioning and sensing services;

- Device diversity: Beyond pedestrian's handheld devices, the network will provide connectivity to a large range of devices mounted on board vehicles, vessels, aircraft, trains, and drones (flying, surface) as well as adapted to fixed installation settings (e.g., utilities' infrastructure);

- Security and privacy: The NTN may reinforce the overall security framework with the distribution of advanced and robust keys to network nodes.

9.3.2 6G Versus 5G Non-Terrestrial Network Component

The 6G definition [5] will have to take into account a general context calling for spectrum scarcity, drastic environmental impact reduction, and high resiliency given that telecommunications are a key component of the global economy. This sets new design challenges for the mobile system for which the NTN component is not just an add-on component providing "connectivity to the unconnected" as in 5G, but a native network component of 6G to support improved service experience, network operation, as well as new use cases, especially for handheld terminals as well as drone/vehicle mounted terminals allowing greater efficiency of human actions while ensuring lower impact on its environment.

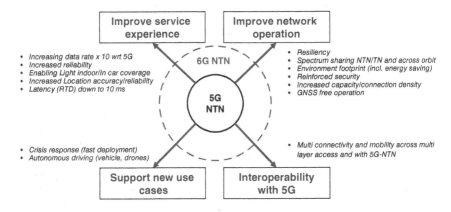

Below, we provide a preliminary list of new use cases that NTN in 6G should support:

- Fast set-up of an autonomous network over a specific region via satellite (with ISL) and/or HAPS with no or intermittent connectivity to core networks (e.g., for crisis response)
- Provision of emergency services (at least SMS) via satellite in light indoor/in car conditions
- Trusted and accurate determination of UE location via satellite networks
- Enhanced connectivity to consumer Handheld (e.g., for video)
- Provision of broadband connectivity to (semi) autonomous cars and drones (including Urban air mobility) and true seamless global service continuity (zero packet loss/zero interruption/no service rate degradation) in high mobility, thanks to NTN/TN combination
- Energy efficient service delivery in multi-access technology network (i.e., NTN/TN)

- Flexible spectrum usage in multi-access technology network (i.e. NTN/TN)
- Hot resiliency with respect to Temporary network node failure in multilayer network (i.e., 3D multi-orbit and meshed network = NGSO, GSO, HAPS, and/or drone-based network node)
- Optimized traffic routing between bidirectional and unidirectional access links

Hence, compared to 5G, NTNs in 6G are prone to achieve 10- or 100-fold performance improvement from 5G to 6G. Preliminary target service performances are provided for indication in the table below:

Target service performances	NTN in 5G (As per 3GPP and/or ITU Radiocommunication sector (ITU-R) IMT2020 satellite specifications)	NTN in 6G
Data rate (DL/UL) wrt Handheld and low-cost IoT devices	1/0.1 Mbps (Outdoor only) @ up to 3 km/h	Outdoor conditions: Tens of Mbps @ up to 250 km/h Light indoor/in car conditions: At least Short Message Service capability
Data rate (DL/UL) wrt Vehicle or drone (flying and surface) mounted devices	[50/25] Mbps @ up to 250 km/h (with 60 cm aperture)	Hundreds of Mbps (Outdoor only) @ up to 250 km/h (with <20 cm equivalent aperture)
Data rate (DL/UL) wrt Large Aeronautic, maritime platforms mounted devices	[50/25] Mbps @ up to 1000 km/h	Thousands of Mbps (Outdoor only) @ up to 1200 km/h (with <60 cm equivalent aperture)
Location service (target accuracy and acquisition time) in outdoor conditions only	respectively 1 m and <100 s (reliability through Network verification)	respectively [0.1] m and <1 s (95% reliability through Network positioning method)
Coverage	Outdoor only	Light indoor/In car
Reliability	Up to 99.99%	Up to 99.999%
Latency	Down to 10 ms (RTD) at 90° elevation	Down to 10 ms (RTD) at 30° elevation
Connection density	Up to 500/km²	>1000/km²

The above mentioned target performances are not consolidated yet and do not preclude what will be adopted in the standard.

Below, we list the frequency bands that may be considered for 5G and 6G NTN, respectively:

Spectrum	NTN in 5G (Currently)	NTN in 6G
Connectivity to smartphones and low-cost IoT devices	FR1: FSS and MSS allocations in L and S bands	FR1: same as 5G-NTN + FSS and MSS allocations in C band
Connectivity to vehicle/drone mounted devices and to large Aeronautic, maritime platforms	Above 10 GHz: FSS and MSS allocations in Ka band	Above 10 GHz: same as 5G-NTN + FSS and MSS allocations in Ku and Q/V bands

In some of these bands, NTN in 6G may have to operate under spectrum sharing with "terrestrial" mobile networks.

9.3.3 6G NTN Design Principles

The following paragraphs infer general design principles stemming from the 6G requirements that are presently shaping up. Key characteristics of an NTN component of the 6G system, and to some extent the beyond 5G system, stated in [6] are graphically represented in Figure 9.6.

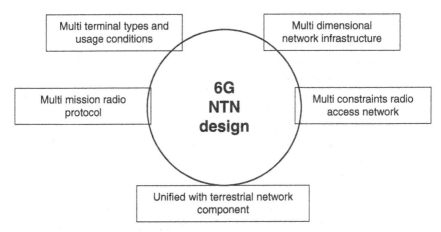

Figure 9.6 Key design principles of non-terrestrial networks for B5G/6G.

9.3.3.1 Multi-terminal Types and Usage Conditions

The strength of an NTN component of 6G lies in its capability to provide connectivity to a wide range of mobile devices (pedestrian, land/air vehicle-mounted, flying/surface drone-mounted, aeronautical or maritime moving platforms, and satellite-mounted), fixed devices (facility-mounted), and ad-hoc networks based on device-to-device communications. Great care shall be taken into account when addressing usage conditions associated with these different devices.

For pedestrian-carried devices, the NTN capability must not significantly impact the design of handheld UE, even in light indoor or vehicle conditions, although the latter could possibly reduce capability or service performance due to building penetration loss.

In vehicle-mounted devices, the automotive industry constraints the size of terminals likely embedded in the roof of the vehicles to maximum 20 cm × 20 cm or smaller (see [7]) and the energy consumption to be similar to that of handheld devices. Furthermore, the industry is called cost-effective antenna capable of tracking the motion of satellites. This framing requires compatibility with the connectivity requirements of the mass market. This will also benefit public safety, media, and entertainment (including news-gathering professionals) and defense applications.

The same installation constraints typically apply to medium-range drones, and railway trains as well as to fixed-mounted devices, especially deployed by utilities and telecom operators.

Even though most aeronautic and maritime vessels may accommodate larger terminals, sticking to small terminals would secure compatibility for all types of vessels and airplanes.

9.3.3.2 Multi-mission Radio Protocol(s)

Leveraging 4G and 5G waveform as well as access layer to the maximum extent is expected, but some enhancements will be needed to support the following new capabilities:

High channel bandwidth flexibility on the radio protocol's waveform is needed to withstand varied terminals and usage conditions with a capability to provide a high range of link margin as well as efficient techniques to mitigate interferences between NTN nodes or between NTN/TN nodes.

Efficient reliability can be implemented by means of re-transmission in a given time period through repetition, interleaving, or power/coding dynamic adaptation.

High-accuracy and trusted localization capabilities can be made available natively by NTN infrastructure, through the exploitation of the motion of the satellite network nodes. This Positioning, Navigation and Timing (PNT) feature would reduce dependency on the GNSS modem embedded in terminals, thereby reducing their cost and consumption while addressing other issues such as privacy.

There is an opportunity to add sensing capabilities to the NTN infrastructure. Leveraging the continuous and global coverage of the infrastructure, high-resolution sensing services may be enabled, thanks to smart signal processing techniques using the 6G radio interface properties. For instance, this can be used for surveillance of the maritime, aeronautic, and possible land transportation traffic.

9.3.3.3 Multi-dimensional Network Infrastructure

It is expected that the NTN infrastructure will be 3D in nature as it should rely on network nodes located on different space orbits, manned or unmanned high altitude and aerial platforms, all interconnecting with horizontal (same-layer) or vertical (across-layer) links.

Network nodes should be providing storage, computational, and routing resources while supporting extensive maintenance and upgrading.

Internode links, whether radio (up to THz) or optical, should offer high service availability and reliability to prevent data loss and link interruption, possibly being reconfigurable to reroute traffic through a different node path upon link failure. In addition, they should be reconfigurable to mitigate the in-plane, inter-plan, or even inter-orbit conditions. The dynamic topology of this network and the variety of services imply a complex orchestration of the resources and virtual functions.

This infrastructure will have to be designed to support the following capabilities or characteristics:

- A 3D NTN infrastructure also supports improved resiliency to natural and man-made disasters, thanks to higher multiplicity of paths connecting network nodes. It will require novel smart routing and multi-connectivity techniques across these multi-paths.
- Always-on: In 5G, NTN are based on payloads, which transparently forward the radio protocol received from the UE (via the service link) to the NTN Gateway (via the feeder link) and vice-versa. Hence, service can only be provided with a serving gateway deployed in the targeted service area. Thanks to the introduction of regenerative payload, UE-to-UE communication will be possible without going through a serving feeder link/gateway deployed in the targeted service area. This will allow to address maritime zones or very remote areas where it is hard to deploy ground network infrastructure.
- Low latency: One of the multidimensional network infrastructure layers may be made of a very low-orbiting constellation of satellites to support

low latency services. In addition, the on-board storage and computational resources will allow to support Multi-access Edge Computing and reduce the end-to-end latency associated with applications embarked on the network node.

Edge computing and storage resources implemented on all the network nodes can be used to optimizing some parameters impacting the service such as the response delay but also the bandwidth of the different inter-node, feeder, and service links. This will require novel security architecture to protect the storage and the computing of user data on board the different flying network nodes interconnected via temporary wireless links to form a dynamic network topology.

As for the impact of space sustainability requirements, they translate into:

- Avoiding the generation of space junk, by operations preventing in-orbit collisions or careful end-of-life disposal of satellites;
- Reducing satellite surface glints possibly disturbing astronomers;
- Minimizing the overall carbon footprint (including the energy consumption) of the network infrastructure production and operation, extending to ground network nodes and user terminals;
- Complying with regulations on user exposure to Electromagnetic Fields (EMF).

9.3.3.4 Multi-constraints Radio Access Network

Artificial Intelligence (AI) will also be needed to always optimize the radio resources of the NTN when addressing the following constraints:
Interference mitigations caused by:

- The sharing of spectrum between NTN and TN. To accommodate the traffic growth of NTN in conjunction with TN, new spectrum bands shall be considered in both sub-6 GHz band and millimeter wave spectrum. Given the scarcity of available spectrum, novel techniques enabling spectrum sharing across mobile services and/or satellite services must be explored. As an example, THz bands may be used indoors for mobile communication and be allowed altogether for inter-satellite links.
- The re-use of spectrum between flying non terrestrial platforms (satellites, HAPs, and drones) at different orbits, which will benefit from spectrum coordination scheme across all orbits.
- The use of smaller size terminal operating in frequency bands above 10 GHz, which may require revisiting spectrum coexistence constraints between adjacent satellites as seen from the terminal.

Limited on-board storage and computing capabilities that require specific management according to the services.

The following aspects shall be taken into account to lower the carbon footprint:

- The minimization of the overall carbon footprint (including the energy consumption) of the network infrastructure, especially the on-ground network nodes and user terminals;
- The compliance with requirements on users' exposure to EMF.

9.3.3.5 Unification With the Terrestrial Network Component

Some enhancements are expected at least in the following aspects:

- Multi connectivity between NTN and TN is a likely feature as well, to support zero packet loss or zero interruption during a data session transitioning between the TN and NTN as UE moves physically.
- Dynamic allocation and routing techniques of the traffic between TN and NTN resources according to the varying spatial and temporal demand.
- Adoption of the foreseen Quantum key distribution-based security framework for 6G, adapted to specific constraints related to the deployment of storage and computational elements in the flying network nodes and the dynamic network topology.

9.3.4 Possible Evolution of NTN Standards

9.3.4.1 Enhancements of NTN 5G-Advanced in 3GPP

During the 3GPP plenaries held in mid-December 2023 in Edinburgh (UK), the scope of Release 19 has been approved. The core part of this Release is expected to be completed in June 2025. This includes several features to further enhance the NTN component in order to improve the service experience, as well as to optimize the network operation in the context of 5G-Advanced system. One should distinguish new features for NR-NTN and IoT-NTN, respectively, [8, 9]:

Enhanced system architecture for the support of satellite access (see 3GPP SP-231199, SP-231802, and TR 23.700), which includes the following main topics in addition to charging, backhaul, and security aspects:

- Support of Store and Forward Satellite operation (e.g., regenerative payload based on eNB with switching and edge computing capability on-board) to support delay-tolerant services in a lacunar (discontinuous coverage) NGSO constellation with a reduced number of gateways.
- Support of UE-Satellite-UE communications for 5GS (i.e., mesh connectivity between UEs via a space segment) supporting NR NTN NGSO

constellations with on-board regenerative payloads and with or without ISLs, with a feeder link always available.

- Dual steer over multiple access links, among which at least one NTN access link. For example, the combination of GSO and NGSO access links or of GSO (downlink only) and TN access links, or of NTN and TN access links will allow to increase the throughput and/or the reliability and/or the quality of experience associated to a service.

Enhanced physical/access layers and architecture of the Next Generation Radio Access Network (see 3GPP RP-234078), with the following further enhancements of NR-NTN in Rel. 19:

- Downlink coverage enhancements, including increased repetition scheme or equivalent techniques on selected physical channels as well as enhanced dynamic and flexible transmit power sharing between active and inactive beams to accommodate reduced EIRP density per beam and to increase the instantaneous service area per satellite(s) that inherently feature limited transmission power and limited processing bandwidth.
- Uplink capacity/throughput enhancement with enhanced multiplexing techniques on the uplink.
- Enhanced signalling the service area of a broadcast service in NTN which may be smaller than the radio beam coverage.
- NTN/TN mobility enhancement to ensure service continuity in connected mode, for example for automotive/transport users that may drive on a road which is intermittently covered by a terrestrial network while fully covered by a non-terrestrial network.
- Regenerative payloads embarking some of the 5G system functions (e.g., gNB and UPF on board) that will primarily allow service in areas without any feeder link (via ISLs), enabling UE-to-UE communications via satellite and improved spectral efficiency on feeder link. The intent is to support services in areas where the NTN gateways are not deployed or temporarily not operational.
- UE with Reduced Capabilities to address low-cost and low-complexity devices supporting wideband services, especially for FR1 through relevant RRM requirements.

Enhanced physical/access layers and architecture of the 4G Radio Access Network (see 3GPP RP-234077); with the following further enhancements of IoT-NTN in Rel. 19:

- Support of regenerative payloads with Store and Forward Satellite operation (e.g., eNB with switching and edge computing capability on-board) to support delay-tolerant services in a lacunar (discontinuous coverage) NGSO constellation with a reduced number of gateways.

- Uplink capacity/throughput enhancement with enhanced multiplexing techniques on the uplink.

Candidate RF and RRM related topics for NTN are expected to be further discussed in March and June 2024. These include:

- UE with enhanced RF performances (e.g., higher sensitivity and/or transmit power) to increase the service rate on both the downlink and the uplink, and/or to extend the service availability in low SNR areas for both NR-NTN and IoT-NTN.
- Vehicle Mounted Relay (VMR) to allow 5G connectivity to access points embarked on moving platforms, such as land vehicles, vessels, or airplanes.
- Definition of new NTN bands, such as Ku and C bands for NR-NTN and an unpaired L-band for IoT-NTN.

Furthermore, some additional NTN related topics may be added in September 2024 in the Rel. 19 workplan, e.g.:

- Notification/alert messages for NR-NTN with enhanced paging channel/procedure to mitigate missed emergency calls or messages in areas with low SNR.
- Support of unpaired spectrum for IoT-NTN.
- 5GC supporting IoT-NTN.

9.3.4.2 Potential Enhancements of NTN for 6G

The following research areas have been discussed [5]:

- Waveform with flexible channel bandwidth and low PAPR to accommodate a wide range of propagation loss (including building penetration) as well as adapted to extend the link margin to address light indoor conditions.
- Radio procedures (e.g., uplink synchronization, initial access) enable GNSS-free operation, especially for terminals with no GNSS capability.
- AI-driven RRM enabling spectrum sharing between NTN and TN.
- Trusted Network-based positioning enhancements for NTN.
- Waveform with integrated sensing and communication capabilities.
- Support of Multi connectivity between NTN nodes as well as between nodes of a TN and of an NTN at RAN level.
- Routing protocols in multidimensional networks with varying inter-node link conditions.

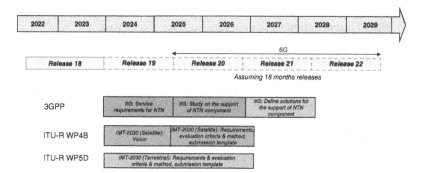

Figure 9.7 3GPP standardization schedule with respect to IMT-2030 definition in ITU-R.

- Enhanced mobility procedures between NTN and TN ensuring zero data loss/zero interruption.
- Split RAN architecture across the feeder link of a satellite network accommodating the limited feeder link bandwidth, the limited processing capability on-board satellite, and the feeder link switchovers
- Enablers for a flexible architecture able to support centralized or distributed network design according to the available edge resources and throughput to connect the edge nodes. This includes relevant security framework.
- Waveform supporting non-orthogonal multiple access for high density of connected devices

In the schedule reported in Figure 9.7, are the initial activities leading to the definition of 6G system in both 3GPP and ITU-R.

References

1 Richard Locke- from Inmarsat "Satellite Communications: Challenges and Opportunities of Future Air Interfaces", ETSI's Workshop on Future Radio Technologies: Air Interfaces, 27–28 January 2016.

2 DRAFT NEW RECOMMENDATION ITU-R M. [IMT.FRAMEWORK FOR 2030 and Beyond] "Framework and overall objectives of the future development of IMT for 2030 and beyond, Document 5/131-E 29", June 2023].

3 ITU-T, FG-NET-2030: "Network 2030 Architecture Framework", June 2020. Available at: https://www.itu.int/en/ITU-T/focusgroups/net2030/Pages/default.aspx.

4 5G-PPP White Paper "European Vision for the 6G Ecosystem", June 2021. Available at: https://5g-ppp.eu/wp-content/uploads/2021/06/WhitePaper-6G-Europe.pdf.

5 Horizon Europe funded 6G-NTN project, 2023. Available at: https://www.6g-ntn.eu/.

6 ESA funded EAGER project (technologies And techniques for satcom beyond 5G nEtwoRks), White Paper, "Architectures, services, and technologies towards 6G NTN", 2023. Available at: https://www.eagerproject.eu/.

7 5GAA position on secure space-based connectivity program and focus on the European communication satellite constellation, 2022. Available at: https://5gaa.org/

8 RWS-230048: "Consideration on RAN1/2/3 led NTN topics for Release 19", Taipei, June 15–16, 2023 at 3GPP Rel-19 workshops, Thales, Hughes, SES, Inmarsat, Ligado, Eutelsat, TTP, Lockheed, Novamint, Airbus, Lockheed Martin, ST Engineering, Sateliot, CeWIT, TNO, JSAT, Gatehouse, Omnispace, ESA, Intelsat, OneWeb, Fraunhofer IIS, Fraunhofer HHI, TNO, IRT Saint Exupery, Hispasat, Gilat, Terrestar, Magister solutions, OQ Technology.

9 RWS-230049: "Consideration on RAN4 led NTN topics for Release 19", Taipei, June 15–16, 2023 at 3GPP Rel-19 workshops, Thales, Hughes, Fraunhofer IIS, Inmarsat, Sateliot, Ligado, Omnispace, Lockheed Martin, Novamint, Eutelsat, TTP, Terrestar, ESA, Intelsat, OneWeb, Airbus, JSAT, TNO, IRT Saint Exupery, Hispasat, Gilat, Gatehouse, Magister solutions, OQ Technology.

Index

Note: *Italicized* and **bold** page numbers refer to figures and tables, respectively.

5G Non-Terrestrial Networks: Technologies, Standards, and System Design, First Edition.
Alessandro Vanelli-Coralli, Nicolas Chuberre, Gino Masini, Alessandro Guidotti, and Mohamed El Jaafari.
© 2024 The Institute of Electrical and Electronics Engineers, Inc. Published 2024 by John Wiley & Sons, Inc.